Your MOBILE Companion

By Roger Burch, WF4N

Published by: **The American Radio Relay League**
225 Main Street, Newington, CT 06111-1491

Contents

Foreword

Hams have been operating from their vehicles since the 1920s, but there's never been as much interest in "goin' mobile" as there is right now. Most hams have chatted with friends, and checked on traffic conditions, over the local 2-meter repeater while commuting to work. But how many of us have worked Asia while mobile—on 20-meter CW? Or tractor mobile while dealing with a stubborn tree stump? Or VHF simplex while flying with a buddy in his Cessna at 4000 feet?

The only limits to mobile operating are some commonsense rules, and our fertile imaginations. Inventive hams have found ways of bringing their favorite hobby with them in any number of ways: over back roads, up the river, cross country on the interstate—and while hang gliding.

If you're looking for a new way to enjoy ham radio, this book will get you going. Chapters cover:
- choosing the right band
- selecting the right gear
- installing transceivers and antennas
- operating, and
- dealing with interference

Whether you already operate mobile, or you want to find out what all the fun is about, you'll find what you're looking for in *Your Mobile Companion*. If you discover (or are already using) a mobile tip or technique you don't find in the book, we'd like to hear from you. Tell us about it on the Feedback form near the back, and we'll consider it for the next edition.

David Sumner, K1ZZ
Executive Vice President
November 1995

Acknowledgments

Countless individuals have contributed ideas and information for this book through many years of QSOs—both on and off the air. To list them all is impossible; to thank them is my sincere pleasure.

In addition, a special word of gratitude to Paul Danzer, N1II, Joel Kleinman, N1BKE and Cary Hill, KA4HHW for their support and assistance, as well as to my good friend J.C. for his unfaltering encouragement.

CHAPTER 1

Going Mobile

Does this sound familiar?

I hate to rush off on you Steve, but I'm in the parking lot here at work, and if I don't get on in, they are going to deduct this conversation from my paycheck. Thanks for riding along with me; it sure makes the trip go faster, 73. KN4EC, clear.

That's perfect timing, D.D. I just pulled into the driveway and I need to get in and get some sleep. Boy do I hate midnight shift! Good morning and 73. KQ4JL clear.

Tune in on any given day to any of the more than 18,000 repeaters around the country, and you are likely to hear a conversation a lot like this one. Besides the fact they both are hams, what else did they have in common? That's right—they were both operating mobile. Nothing unusual about hams combining their favorite hobby with our "love affair with the automobile," right? When you consider the average American motorist drives more than 12,000 miles per year, it's only logical those motorists who are hams would take along a radio of some sort. Or in other words, "go mobile."

And that's what this book is all about: Mobile operation. From cover to cover, we'll be looking at the nuts and bolts of mobile operation. We'll look at who operates mobile, and why. We'll take you from selecting what bands to operate to selecting the radio equipment you'll use. We'll show you how to install your equipment, and how not to. We'll help you with antenna selection and installation as well as the routing of cables. We'll look at automotive electronic systems and how you can help your radio equipment coexist peacefully with them. And we'll be looking at lots of ways to operate mobile, including some of the more unusual methods of mobile operation. But before we get down to who, what, when, where and how, let's look at a little bit of the history of mobile ham radio.

Back in the Good Old Days

"Did he say history? What history?" you may be asking.

Perhaps you are thinking the practice of driving around with a radio in the car and talking to other hams has only recently become popular. In fact, hams have been taking their radios to the streets since our hobby was in its infancy. There's even a rumor floating around, unconfirmed of course, that before radio licenses were required, Henry Ford used to roam the streets of Detroit in his Model T, working spark-gap CW with a spare ignition coil and a trailing wire antenna!

The accompanying photos show a mobile station from the early 1950s. Although some VHF mobile operation was taking place at the time, it was, for the most part, limited to metropolitan areas. The absence of repeaters on the VHF bands dictated HF operation for communication over any appreciable distance.

As Figs 1-1 and 1-2 show, the ham of yesteryear may

Fig 1-1—Photo of the "serious" engine modifications for W6ZV's 75-meter mobile installation. An additional generator, carbon-pile voltage regulator and reverse-current relay combined with hash filters were all needed for this state-of-the-art installation. The picture was published in January 1952 *QST*.

Fig 1-2—W6ZV's vehicle, as it looked as he tooled around 6-land in the early 1950s. The original caption read: "The high-power mobile antenna of W6ZV is a potent putter-outer and a real attention-getter."

have sported nearly a hundred pounds of radio gear stuffed under the dash of his Hudson Hornet, and another couple of hundred pounds of gear in the trunk. Yes, those tubes, transformers and dynamotors were pretty heavy–quite bulky, too! The January 1952 *QST* article included photos describing in detail how W6ZV put together a 1000-watt mobile installation for 75 meters, complete with a dual 10-foot long whip antenna system, engine-mounted 28-volt 150-amp generator and a trunk full of kW amplifier.

W6ZV was not alone in his efforts to free Amateur Radio from the shackles of operating from a fixed location.

Fig 1-3—The ARRL membership ad from January 1952 *QST* trumpeted the burgeoning activity of mobile operating.

In the same issue of *QST*, a photo ad promoting ARRL membership featured a "trunk-mounted job with a driver-compartment control" (see Fig 1-3).

The lifting of the wartime ban on Amateur Radio transmissions, as well as the threat of nuclear attack, had regenerated interest in mobile operation. As Fig 1-4 shows, mobile operating had become popular well before WW II.

Why Me?

Surely by now you are convinced that the Biblical observation "there is no new thing under the sun" certainly applies when it comes to mobile hamming. But have you ever given serious thought to why we operate our radios from cars, trucks, vans, boats, planes, motorcycles, bicycles, buses, trains and a host of other forms of transportation? Okay, I can hear some of you seasoned "Road Warriors" out there chuckling to yourselves, "'cause it's con-

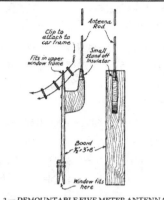

5-Meter Antenna for the Car

Needing a method of supporting a 56-mc. vertical antenna on a car so that the antenna would be readily demountable and not mar the car in any way, I devised the scheme diagrammed in Fig. 2. It worked very successfully, so I am passing it on to any of the five-meter gang who may be in the same predicament. The car windo is cranked all the way donw, thupper (sharpened) end of the flat wood piece is fitted in the groove where the glass usually goes, and the window is then cranked up into the groove in the lower end o fthe wooed piece. For the antenna itself, a lathe or bamboo flower stake supporting a wire would work just as well as the rod and stand-off combination, which is not very strong in a wind. I use a 46-inch rod with six-foot feeders space two inches, a combination which works out well for mobile operatinon where trees are lowe over the streets. A small battery clip grounds the odd feeder to the metal rain gutter around the roodf of the car.

—*Edgar V. Seeler, Jr., W3BBZ, W1BDF*

FIG. 2 — DEMOUNTABLE FIVE-METER ANTENNA FOR THE CAR

Fig 1-4—This intriguing item appeared in the For the Experimenter column in October 1934 *QST*. The 5-meter band was the predecessor to today's 6-meter band.

venient!," "it's fun!" or, "everyone's doing it!" Well . . . all those arguments actually are quite valid. Mobile operation is convenient and it's fun, too. And it does seem just about every ham *is* doing it. But perhaps there *is* more to this concept of operating mobile than meets the eye.

Rollin' Down the Highway

It's Saturday evening of the eighth day of my long-awaited vacation. As I slowly cruise the streets of a coastal town in Southern California, I watch as the noise relentlessly keeps the S meter on my HF rig peaked near S9. Having logged more than 600 miles earlier in the day, the last thing I want to do is to get out and drive around some more. That's just what I'm doing, however, hoping to escape the noise threatening to prevent me from keeping my schedule with Dale, NU4O, my neighbor back in Kentucky. I'm hoping if I get away from our motel parking lot, so generously lighted with high power mercury vapor lamps, it will solve my static problems. Apparently, the nearby power-generating station and the even closer substation are the real culprits. Realizing the time for my schedule with Dale is just minutes away, I turn onto a beach access road and begin to look for a place to park, all the while hoping Dale's signal will be strong enough to beat out the noise.

Backing into a pull-off just a few yards from the Pacific Ocean, I shut the car engine off and crank up the volume just a bit. As I slowly tune my rig up and down in frequency near 14.020 MHz, suddenly through the noise comes

WF4N/M6 WF4N/M6 de NU4O NU4O BK

Startled, but certainly relieved to see Dale's signal nearly 20 over 9 on the S meter, I increase the speed on my keyer just slightly and return,

NU4O de WF4N/M6 FB your signals very good here. How goes everything? BK

Everything here FB too. Your signals very good for a mobile. S9 with a little noise but very solid copy. BK

FB Dale, maybe the ocean is providing a good ground plane for me. How is everything back there? BK

Everything here is good except the weather. By the way, I sure do appreciate you keeping up with us on this trip. I can not tell you how good it is to be in contact everyday with someone back there at home. BK

No problem Dale, I know I have kept you up late every night this week. Again, I sure do appreciate you keeping this sked with me. 73 and gn de WF4N/M6 SK

Take care and see you tomorrow morning. 73 de NU4O SK

Many hams may never try mobile CW. But that's not

what this actual replay of a QSO was all about. The essence of what Dale and I were doing was *keeping in touch*. That's all. It's something any ham who has mobile capability can do, and it's one of the most popular reasons hams operate mobile. It doesn't matter whether we're going around the corner or around the world—across town or across the continent. Nor does it matter whether we are staying in contact with a spouse, child, parent, friend or whoever happens to show up on the frequency; having a radio along when we are on the go is our umbilical cord to the rest of the world. Granted, modern technology (cellular phones, for example) has made it possible for traveling non-hams as well as hams to stay in touch. Although a cell phone can be quicker and more convenient, especially if your need is to talk to someone who isn't a ham, if you like to ragchew for hours on end as you travel across the country, please don't send me the phone bill!

Many a traveling ham has been kept awake, informed or entertained by a fellow ham at the other end of the "ether canal." As those who do it will quickly attest, it's one of the best reasons in the world to operate mobile. But there are many other good reasons to operate mobile.

Competitive Mobiling: The Heat is On

Are you the competitive sort? Does a friendly game of one-upmanship strike your fancy? If so, there is a good possibility you may already have tried your hand at contesting. There's a ham radio contest of some sort just about every weekend throughout the year on one or more of the amateur bands. On some weekends, you may find several contests in progress, each one pitting hundreds or thousands of hams against each other, all vying for the coveted *Numero Uno* position in the score box. Since contests generate so much interest, each month *QST* devotes an entire column to upcoming contest rules and other operating information.

So why not try mobile contesting. If you've contested, but not while mobile, you're probably asking what possible reason there could be to operate mobile during a contest. After all, isn't it pretty demanding to just stay in the thick of a contest, much less trying to contest and drive at the same time? And besides, you don't get any extra points for operating mobile, do you?

Not necessarily. Actually, there *are* some contests that do allow extra points for contacts made while mobile. Check the contest column of your favorite magazine to find those that reward mobile operators with extra points or multipliers. They usually include, but aren't limited to, statewide contests.

Another reason for mobile contesting, and a very popular one at that, is that some contests effectively encourage mobile operation by allowing you to count multiple contacts with the same station as long as one of you changes location for each contact. Generally, on the HF bands this will require you to operate from

Fig 1-5—Compare this photo of Robert Craft, W6FAH's mobile to the one in Fig 1-2. Even with two antennas, this installation would be very inconspicuous, except for its location: on a bright red Corvette! *(Photo courtesy of W6FAH)*

a different county in a given state for each same-station contact to count. On the VHF and higher bands, a system of *grid squares* (we'll be discussing contests in detail later) is used to determine your location, with each grid square you operate from allowing you to count same-station contacts the same as contacts with new stations. It doesn't take a lot of imagination to see the implication here: The mobile ham has the advantage of no longer being the hunter; he or she is now the hunted! Each time you enter a new county or grid square, you again become a "good contact"—just what every contester wants! It's like stocking your swimming pool with large-mouth bass, then firing up the smoker, pulling up a lawn chair to the edge of the water and going fishing. All you have to do is throw out the bait (your CQ and your new location), and reel in the fish (new contacts). You just *thought* you had to mount a DXpedition to be on the receiving end of a pileup.

DX—A Little Extra Spice!

Oh, and speaking of DX, mobile operation works curiously the same way when you are chasing DX on the HF bands. While mobile whip antennas and barefoot rigs can't easily compete with the kilowatt stations and their beam antennas when it comes to chasing DX, it is surprising how many DX stations will go the "extra mile" to work a ham who is signing mobile. The mobile operator seems to enjoy a sort of underdog status in the heart of many DX operators—something you can certainly use to your advantage when you are trying to work a "rare one."

Filling the Need

So far, we have looked at the casual or entertaining reasons for operating mobile. There is certainly nothing wrong with operating mobile strictly for the pleasure of doing so. After all, that's why most hams do it.

Now let's shift gears and look at one very compelling reason every Amateur Radio operator *should* have some means of mobile operating capability. It's called *public service.* Granted, this term can encompass many activities, including message handling through the National Traffic System and providing tactical communication at parades, marathons and other public events.

Perhaps the most important aspect of public service communication, however, is furnishing communication during times of emergency or disaster. That's when mobile operators really shine. Sometimes traveling hundreds of miles to make their communications services available to an area stricken by disaster, they are the unsung heroes to many folks who have been left helpless, hurting and without communication. So if you're the type of person who likes to help others (it seems to come with the ticket), you'll find the ability to operate mobile is an excellent way to funnel your "helping hands" energies. For example, if you live anywhere near "Tornado Alley," you can be a valuable asset to local authorities by volunteering as a weather spotter— someone who goes to a location with a good vantage point and watches for the approach of severe weather. Or you may be dispatched to an area devastated by a tornado, where you might be the first person to provide damage reports. On the other hand, you may live in an area prone to earthquakes. Again, your ability to get out and be on the scene of a disaster soon after it occurs can be of immeasurable worth to the victims—who may find you to be their only link to the outside world. Fact is, no matter where you live, the possibility of a disaster occurring, naturally or otherwise, always exists. If it does occur, you'll never regret being able to provide essential communications services to those who need them. Having a well-equipped mobile station is the first step to being prepared to meet this need.

But what if you don't live on the San Andreas Fault or in Dorothy and Toto's neighborhood? Does that make it

Get Involved

Most areas have a volunteer known as the *Emergency Coordinator*, or EC. The EC is part of the ARRL *Field Organization*—individuals who serve the community and their fellow hams by organizing local hams to be responsive in times of emergencies. You can contact your EC by calling or writing your local *Section Manager*, or SM. The address and phone number of your SM is printed each month on page 8 of *QST*.

Years ago, only hams had the radios that let police, fire, Red Cross and other emergency services talk and coordinate activities with each other. Back then, a "modern" town or county may have had one *hotline* frequency and a few radios that could be switched to this frequency, but generally the radios and manpower to cross organization boundaries came from hams.

Today, it appears that everyone carries a hand-held radio. Trunked systems on 800 MHz are replacing the VHF/UHF systems previously used, and coordination is the name of the game with trunked systems. So why do hams still play an important role?

Emergencies are, by definition, unexpected. All sorts of surprises occur and trained people are needed—those who can support themselves with equipment, spares and the know-how to respond quickly. Need a communications point at a school?—send a ham! Power out at a location and you need to talk to someone there?—send a ham! Buildings in the way and normal radio can't be heard in the area?—see if the hams (with a portable repeater) can help.

Flexibility, quick response and trained people—this is the reputation of hams in many communities. You, your mobile and your hand-held radio are a significant part of this service.

unlikely your mobile communications skills will ever be called into action? Hardly! Each year, hundreds of mobile hams provide emergency communications by alerting authorities to the scene of an accident or potential hazard. In fact it's the most common form of emergency communication hams provide. You might be the person who calls a service truck for a mom and her kids who have spent the past hour shivering in a freezing car, broken down on the side of the road. Or maybe you'll be the first to report a road hazard to the state police and save some traveler from receiving serious damage to their vehicle. Or perhaps you'll suddenly find someone's life dependent on you and your radio, as one ham did.

When Minutes Count

It was just past 11 PM, and I was talking on the local 2-meter repeater to Eddie, WA4EGQ, as he was driving home from work. We were just chatting about the usual stuff. He had just begun to tell me how his evening had gone when he suddenly paused in mid-sentence–then quickly said:

> You know, I'm not sure, but I thought I could see a car coming toward me on the far end of this long stretch of road just outside town, but it just suddenly disappeared. I'm going to have a look just as soon as I can get up there.

I replied:

> Okay, Eddie, be careful; it may be a drunk with his lights off. Let me know what you find.

Just seconds later, he came back:

> It *was* a car and it went down the steep embankment on the south side of the highway. It's so far down there I can't see it very well, but I hear someone calling for help. While you call 911 and get a squad headed this way, I'm going down to check it out. I'll be back in a minute to let you know what I have.

I'm dialing right now and I'll be listening for you.

When the 911 dispatcher answered, I gave him all the information I had. I then told him I would be calling him back with additional information as soon as I received it. He advised me he was sending EMS, and then we both hung up.

After what seemed forever, but was actually just a couple of minutes, Eddie was back on the air, and the urgency in his voice was profound.

Call 911 back and tell them there is one person with injuries and she's pinned in the car. I can't get her out. The car and the area around it all reek of gasoline; the car may go up in flames any minute, so tell them to get a pumper here as quickly as they can. The girl is really hysterical, so I'm going back down to try to calm her a little. Stand by on the frequency for me. WA4EGQ.

Okay, Eddie, I have the dispatcher on the phone now. WF4N standing by.

After about 10 minutes, Eddie was back:

WA4EGQ, mobile again. You still there, Roger?

WF4N still here, Eddie. How goes it?

The pumper and the rescue squad are both on the scene. They've got the area hosed down, so there's no danger of fire now. When I left, they were getting ready to open the car up and get the girl out.

Boy, that's pretty quick. Did you know it's only been about 15 minutes since she went off the embankment?

Well to tell you the truth, it seemed like an eternity to me. But you know what? Events like this sure make me glad I've got a radio in the truck.

While no one can say for certain how critical those minutes were to the rescue of that young lady, and her subsequent full recovery, at least a couple of hams sure were glad a radio was nearby!

Okay, so you've got your motor running and you're ready to head out on the highway and do some public service work. Good! You'll have lots of company; there's nothing many hams like better than to get out and show off their communications skills. But are you aware hams are actually *expected* to provide public service communications? I can hear you now: "Nope, looked over my ticket and *The FCC Rule Book*, and they don't say anything at all about me being required to provide public service communications!"

Well, you're absolutely right. We *aren't* required. But it is expected. "By whom?" you ask. By the FCC! The FCC has always expected Amateur Radio operators to provide public service communications. And realistically, we hams owe the very existence of our hobby to our long and notable history of providing communications when no one else could. Our ability to do so contributes to justifying our frequency allocations.

While some may look at Amateur Radio as just a hobby, the agencies that oversee the allocation and usage of our radio frequency spectrum see it much differently. They see Amateur Radio as a source of skilled communicators who are trained and prepared to step forward when needed to fill the gaps in the emergency communications systems, be it local or nationwide. And this is a good thing for us. In just the slice of radio spectrum extending from our 3.5-MHz band to the upper end of our 440-MHz band, we have more than 40 MHz of radio spectrum—much of it prime RF real estate. And it's high-priced real estate at that! We are now watching as the privilege to use radio frequencies is being sold to the highest bidder, sometimes for prices effectively approaching $800 million per megahertz. And since megahertzes aren't being made anymore, it's a safe bet prices will go up. As prices go up, so will the competition for any and all available spectrum.

Today, more than ever, hams must be ready to prove

our frequency allocations are justified *and* being put to good use for more than just our amusement. Naturally, one excellent way to do it is to have a well-equipped mobile station, ready to hit the road whenever it's needed. In the meantime, of course, it'll make for some very enjoyable hamming!

Let's Go!

Okay, so you're brimming with enthusiasm. You've got the old battlewagon cleaned up, tuned up, gassed up and shod with new rubber, ready to become a superstation on wheels. So where do you go from here? There's a lot to consider when it comes to mobile operation: Where to mount radios and antennas, what bands to operate and a host of other concerns. In the following pages, we'll be looking at those and several other aspects of mobile hamming.

CHAPTER 2

What Bands?

ow that you are convinced "ham-mobility" is for you, what is your next step? Obviously, before you purchase or install a mobile radio, you have to decide what band (or bands) you will be operating. Sound easy? Well. . . perhaps. If you limit yourself to using commercially available equipment, it is still possible to operate mobile on a total of 14 different Amateur Radio frequency bands spanning the range from 1.8 to 1300 MHz.

With such a vast expanse of radio spectrum at your disposal, your choice of bands may seem overwhelming. It need not be—with a careful evaluation of what each band has to offer, you will be sure to find your niche in "mobile hamdom."

Before getting into a detailed examination of the various bands, let's look in a more general sense at what makes a band attractive to the mobile operator.

Is Someone There?

Here's an easy one: What is the purpose of being a ham

radio operator? To talk to other hams, of course. So how does that affect your choice of band(s) for mobile operation? To start with, you are going to want to choose a band with plenty of activity. You will find this choice to be even more crucial to the mobile VHF or UHF operator, whose range is typically less than a ham operating from home on the same band. Keep in mind, too, that sparsely populated bands are not limited to the spectrum above 900 MHz. Many things affect the popularity of an amateur band. Local convention, the number of hams in a given area—even the sunspot cycle—are all factors in *who* operates *where*.

Of course, if you are already active on one or more bands from your home station, this might be a persuasive argument for assembling a similar setup for mobile operation. After all, you will already be familiar with the characteristics of these bands as well as knowing many of the hams who frequent them. In addition to not having to "plow new ground," operating mobile on the same bands as you do from home can allow the use of the same rig for both locations. You will, of course, have the added chore of transferring the equipment to and from your vehicle.

But what if you are not yet active on any of the amateur bands? Furthermore, you would like to make sure your debut as a mobile operator will not play to an empty house. Check with your Elmer (the ham who introduced you to Amateur Radio) or the members of your local ham radio club to find out what bands *they* are using for mobile operation.

Be careful, however, not to rely strictly on the opinions of others. Hams are a diverse group. You may find what appeals to other hams may not interest you at all. So once you have gotten some valuable input from your ham friends, do a little detective work on your own.

If you are considering the bands above 30 MHz, it is likely you will want to center much of your mobile opera-

Repeaters: Lifeline for the Mobile Ham

Nearly all hams have used repeaters. Why are they so popular? There are many possible answers, but the thing you will like best about repeaters is that they greatly increase the range of your radio.

Usually installed at a high location—on a tall building or perhaps on a mountain or hilltop—the repeater can hear and be heard for a much greater distance than stations operating point to point. What does that mean to you, the mobile operator? By relaying your signal, the repeater can make it possible for you to talk to someone who wouldn't normally be able to hear you. You might simply be too far away, or there might be an obstruction preventing your signal from reaching them. Either way, the repeater "steps in" to pass the signals back and forth.

With such mystical range-multiplying capability, repeaters must surely be super-complicated, right? Not necessarily!

What Do Repeaters Actually Do?

In its simplest form, a repeater has but one function: To receive a signal and retransmit it.

Picture it like this. You take one of your kid's toy space ranger radios, tape the push-to-talk switch down and set it in front of your stereo system's speakers. You then take the other space ranger radio out to the garage where you can now be kept entertained—thanks to your home-made repeater! That's pretty much how an Amateur Radio repeater works.

In a basic repeater, the receiver has its audio output fed into the transmitter, which simultaneously retransmits whatever the receiver is hearing. (Technically, it's more correct to say the repeater is simultaneously transmitting your signal, but the name "repeater" sounds better than "simultaneous-transmitter"!). Since the repeater transmitter doesn't have a microphone, we can't tape the push-to-talk button down. Besides, we don't want the transmitter to be on all the time—just when there's someone using the repeater. So how does the transmitter know when to transmit? Again, in the most basic repeater, a carrier-sensing circuit is used to key the transmitter any time it detects a signal in the receiver. That done, your melodious voice that's arriving at the receiver is being squirted back onto the airwaves by the transmitter.

But wait. If the transmitter and receiver are both operating at the same time, isn't the repeater hearing its own signal? It would be—if it weren't for some basic facts of repeater life.

One of those facts is something called *offset*. This simply means the repeater has its receiver tuned to a different frequency than the transmitter. Let's say, for example, a repeater transmits on 146.64 MHz. This means its receiver is tuned to 146.04 MHz, an offset of 600 kHz (standard for 2-meter repeaters). You'll find different offsets for different bands: 440 MHz uses 5 MHz, on 222 MHz it's 1.6 MHz, 6 meters uses 1 MHz and on 10 meters it's 100 kHz.

Making a Contact

Let's suppose you want to use our theoretical 146.04/.64 repeater to give your friend a call. First, you would set your radio for 146.64 MHz. Then, unless your radio does so automatically, select a negative offset. When you key your radio, it will switch automatically to a transmit frequency of 146.04 MHz, where the repeater will receive your signal and retransmit it on 146.64 MHz. When you've finished talking and un-key, your radio will switch back to its receive frequency of 146.64 MHz. You will now be able to hear the repeater transmitter—and your friend, who's answering you.

Having the receiver and transmitter frequencies offset

is necessary to keep a repeater from talking to itself, but it isn't enough! Even with separate transmit and receive antennas, early repeater operators found the signal from the transmitter could easily desensitize, or "desense" the receiver, making it impossible for all but the strongest signals to make it into the machine! Unsolvable dilemma? It might be if not for a device called a *duplexer*. Connected three ways between the repeater's receiver, transmitter and antenna, a duplexer is a filter that's so selective, one of its two sections can let a signal going to the receiver pass through virtually unaffected while completely blocking the signal from the transmitter. At the same time, the other section is allowing almost all the power from the transmitter to pass through, while stripping off any spurious energy that might be present on the receiver frequency. The duplexer is so efficient at what it does that the transmitter and receiver can share a common antenna. That's something you would never want to try without a duplexer!

Okay, take one transmitter, one receiver and one duplexer, tune carefully and you have . . . yes—a repeater. Not a very exciting repeater, but it's still a repeater. To spice up their machines a bit, many repeater owners add another ingredient called a *controller*. Essentially a self-contained computer, a controller takes full command over the repeater. It keys the transmitter, provides the tone known as a *courtesy beep*, connects the repeater to the phone line when you want to make an *autopatch* call, transmits the repeater call sign at proper intervals—and it may even talk!

Although sophisticated controllers aren't required for repeaters, everyone agrees they make repeater use much more enjoyable.

tion around one or more repeaters. (To learn more about how repeaters work, see the accompanying sidebar.) If this is the case, the latest copy of the *ARRL Repeater Directory* will be a valuable asset as you map out repeaters of interest in the areas where you will be mobiling. The *Directory* provides you with the locations, call signs and frequencies of all the repeaters in the US. It also furnishes pertinent information on each repeater, such as whether an autopatch is available—or if it is necessary for your rig to transmit a subaudible tone to access the machine. Indispensable to the traveling ham, the *ARRL Repeater Directory* also comes in electronic form for use with a personal computer. *The North American Repeater Atlas*, also available from the ARRL, shows repeater locations drawn on maps. It covers the US, Central America and Canada.

Making Direct Connections

By the way, keep in mind not all VHF/UHF activity occurs on repeaters. In areas where repeaters are often busy, many hams use *simplex* frequencies whenever possible. Going simplex helps relieve the congestion on busy repeaters (and makes the repeater available to other hams who need the communication range a repeater provides).

Since many hams hang out on non-repeater frequencies, and since the presence of a repeater in your area does not guarantee it sees a lot of use, there is no substitute for doing some *listening*.

You probably will not have radios for every band you will be considering, so one of the best ways to scout for activity is with a programmable scanner. If you own or can borrow a scanner, and it can search a programmable range of frequencies, you can use the search feature to quickly survey a band from edge to edge. Once you have easily pinpointed those frequencies popular with the hams in your

Table 2-1

US Amateur Bands

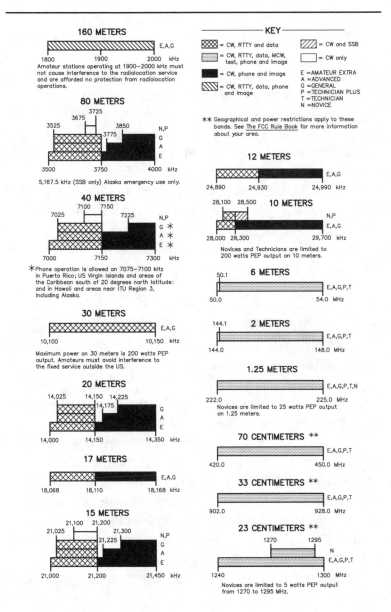

160 METERS

E,A,G

1800   1900   2000   kHz

Amateur stations operating at 1900–2000 kHz must not cause interference to the radiolocation service and are afforded no protection from radiolocation operations.

80 METERS

3725
3675
3525   3850
            3775

N,P
G
A
E

3500   3750   4000   kHz

5,167.5 kHz (SSB only) Alaska emergency use only.

40 METERS

7100 7150
7025   7225

N,P
G ✳
A ✳
E ✳

7000   7150   7300   kHz

✳Phone operation is allowed on 7075–7100 kHz in Puerto Rico; US Virgin Islands and areas of the Caribbean south of 20 degrees north latitude; and in Hawaii and areas near ITU Region 3, including Alaska.

30 METERS

E,A,G

10,100   10,150   kHz

Maximum power on 30 meters is 200 watts PEP output. Amateurs must avoid interference to the fixed service outside the US.

20 METERS

14,025 14,150 14,225
        14,175

G
A
E

14,000   14,150   14,350   kHz

17 METERS

E,A,G

18,068   18,110   18,168   kHz

15 METERS

21,100 21,200
21,025   21,300
        21,225

N,P
G
A
E

21,000   21,200   21,450   kHz

KEY

= CW, RTTY and data

= CW, RTTY, data, MCW, test, phone and image

= CW, phone and image

= CW, RTTY, data, phone and image

= CW and SSB

= CW only

E =AMATEUR EXTRA
A =ADVANCED
G =GENERAL
P =TECHNICIAN PLUS
T =TECHNICIAN
N =NOVICE

✳✳ Geographical and power restrictions apply to these bands. See The FCC Rule Book for more information about your area.

12 METERS

E,A,G

24,890   24,930   24,990   kHz

10 METERS

28,100 28,500

N,P
E,A,G

28,000 28,300   29,700   kHz

Novices and Technicians are limited to 200 watts PEP output on 10 meters.

6 METERS

50.1

E,A,G,P,T

50.0   54.0   MHz

2 METERS

144.1

E,A,G,P,T

144.0   148.0   MHz

1.25 METERS

E,A,G,P,T,N

222.0   225.0   MHz

Novices are limited to 25 watts PEP output on 1.25 meters.

70 CENTIMETERS ✳✳

E,A,G,P,T

420.0   450.0   MHz

33 CENTIMETERS ✳✳

E,A,G,P,T

902.0   928.0   MHz

23 CENTIMETERS ✳✳

1270   1295

N
E,A,G,P,T

1240   1300   MHz

Novices are limited to 5 watts PEP output from 1270 to 1295 MHz.

area, you will be able to decide which frequencies are of interest to *you* when you are mobile.

And the Winner Is . . .

Okay, suppose you have graduated from the Sherlock Holmes School of Frequency Sleuths, you have done some intense detective work, and you have found plenty of hams in your area are on practically every band. So what do you do? Equip a mobile station that can operate from dc to daylight? If you have the means to do so, you will certainly be the envy of all the hams in your neighborhood. You will also find yourself much in demand anytime there is a need in your area for emergency communication.

For most hams, however, the constraints of budget and limited vehicle space will dictate a more conservative approach. Realistically, your choices will likely be governed by the distance you want to be able to communicate and by the type of operating you plan to do.

Above 30 MHz

The VHF and UHF bands are far and away the most popular for mobile hams. It is easy to see why. The equipment is compact and easy to mount, as are the antennas. In addition, there are lots of other hams to talk to, no matter where you may travel. What is really nice, though, is that communication on these bands is predominately local—and very reliable.

Want to check in with the local "Early Risers Group" every morning as you head to work? If they hang out on one of the VHF or UHF repeaters, you do not have to worry about whether band conditions will be good enough for you and the group to hear each other. If there are other hams in your family, you can catch up on all the last-minute family plans for the day you forgot to discuss before you left for work.

Offering reliable local communications and a large

pool of available operators makes the VHF and UHF bands a natural for emergency use as well. While a major disaster will sometimes require long range communication, the overwhelming majority of emergency communications are local and take place on the VHF/UHF bands. Mobile hams are ideally suited for local emergency communications.

Let's have a look at some of the appealing characteristics of bands over 30 MHz.

6 Meters

Sandwiched between the 30 to 50 MHz business band and the broadcast television Channel 2, the 6-meter band extends from 50 to 54 MHz. Once considered a "no ham's land" because of the propensity of 6-meter transmitters to cause interference to neighborhood TV sets, 6 meters has seen a resurgence in activity in recent years. Why? There are several reasons.

Give Me Some Room

One reason is that 6 meters is not usually crowded. If you don't enjoy waiting to use a frequency, reminiscent of a telephone party line, you may find 6 meters a welcome respite. Of course this does not mean the band is desolate. Many metropolitan areas, as well as some rural locations, have 6-meter repeaters. And as the other bands grow much more crowded, you are sure to see an increase in activity on 6.

CQ DX

If you like to work DX, you will be glad to hear 6 meters experiences more frequent periods of exceptional propagation—and over longer distances—than any other VHF or UHF band. The reliable range during normal conditions on 6 meters might be 30 to 50 miles for an FM mobile working a base station or a repeater. With the push provided by band openings, your 6-meter mobile signal will be transported many times that distance, however. Does the idea of driving around town as you talk to other hams who may be 1500 miles—or more—away sound exciting? The 6-meter band is the only band above 30 MHz where it is going to happen.

Cable TV

Probably the greatest blessing ever to be bestowed on the 6-meter band, cable TV provides a closed system of shielded transmission lines and remotely located antennas and amplifier systems. There is no doubt cable has helped eliminate the TVI curse from the prospective 6-meter operator. This has encouraged many new hams, as well as some old-timers, to visit this formerly forsaken band.

Repeaters, occasional long distance propagation, a growing number of users and some really neat mobile gear—all good reasons to try 6 meters as *your* mobile band.

2 Meters: "Where the Action Is"

When '60s rocker Freddy Cannon was high on the charts with a song by that title, he was not singing about 2 meters. But today you would be hard-pressed to find a more fitting description of our most popular band.

Nestled into a 4-MHz wide chunk of VHF spectrum spanning 144 to 148 MHz, 2 meters is a bustling beehive of activity, offering a smorgasbord-like variety to the mobile ham.

Expect reliable mobile-to-base simplex range on this band to be approximately 20 to 40 miles under normal conditions, although periods of enhanced propagation may increase the range to several hundred miles.

Although 2-meter mobile simplex operation is common, one of the most appealing aspects of the band is the presence of repeaters—*thousands* of them. There is hardly a place in the US where the mobile ham might travel and not be within range of a 2-meter repeater. In fact, 2 meters was the birthplace of the Amateur Radio FM repeater. As 2-meter FM grew ever more popular in the 1960s, hams discovered that a strategically located repeater could give their mobile rigs the same range as a well-equipped fixed station.

As 2-meter activity has evolved over the years, so too has the complexity of repeater systems—fueled by the ingenuity of the hams who construct them. With a feature known as *repeater linking*, you are no longer limited to a range of perhaps 100 miles when working through a 2-meter repeater.

Suppose you are driving from Evansville, Indiana, to Chicago, and you would like to stay in touch with your ham spouse. No problem. By using a system of linked repeaters, the signal from your 2-meter rig will be relayed back to a repeater within range of your home. Your spouse will hear you as though you are next door, even though you are hundreds of miles away.

Of course, linking is not limited to just connecting one 2-meter repeater to another. Sometimes you will find a 2-meter repeater with a *crossband* link, enabling you to have your signal retransmitted on another band. If the crosslinked band is 10 meters, with good band conditions you may find yourself talking to someone halfway around the world!

Autopatches are an indispensable feature to the mobile ham—and you will find them available on many 2-meter repeaters. Simply an electronic interface between the repeater and a telephone line, autopatches make it possible for you to use your radio to call someone on the telephone. It works much the same way as a car phone.

Stuck in traffic as you head home from work? A quick

Repeaters: The Good and the Not so Good

Repeaters are found on all bands from 10 meters on up, but their real popularity is on the 2-meter and 440-MHz bands. The quiet in the car (due to the *squelch* control), increased driving safety with channelized operation and the fact that you can stay in touch with your local group are all plus features. Most stations within range of the repeater are Q5—armchair copy! But if you have never operated on this mode, you'll want to prepare for some challenges.

On long trips, every 40 or 50 miles, depending on the number of hills and mountains, you will have to look up a new repeater frequency and switch to it. During the day, many repeaters have very little activity. Don't be surprised to find repeaters occupied by only a handful of hams, and these only during rush hours.

Other repeaters, in metropolitan areas, are constantly busy—perhaps too busy for you to have a leisurely chat. In addition, one station accidentally (or otherwise) keying the repeater can keep everyone else from using it.

Despite these challenges, repeaters are the most popular means of ham communication today. So jump right in—it's the best way to keep in touch with friends and family, and to make new friends.

call using the repeater autopatch will allow you to inform your family they'd better start dinner without you.

Lots of repeaters, autopatches, plenty of activity—it is easy to see why 2 meters is the band of choice for thousands of mobile hams. So come join the fun.

222 MHz

Want to escape the VHF rat race? Then 222 just may be your ticket to Serenityville. Situated just above the top of the VHF broadcast television band, it resides in a 3-MHz slot extending from 222 to 225 MHz.

Increased crowding in the other VHF/UHF bands has led to increased interest in 222 MHz. Offering range comparable to the 2-meter band, 222 MHz is a popular alternative.

Although you will not find a bewildering maze of repeaters on 222 MHz, there are still many to choose from, with more being brought on-line each year.

If you are a Novice, that's especially good news for you, since 222 MHz is the only VHF band available for your use. The good news gets even better with the knowledge that some 222-MHz repeaters are linked to repeaters on other bands. This gives you the ability to talk on frequencies where you do not normally have operating privileges. Depending on what band is linked to the 222-MHz repeater you are using, it may be possible for you to join the conversations on the local 2-meter repeater. Or, you may even be able to work some DX on 10-meter FM.

As with 2 meters, you will find a good assortment of 222-MHz mobile gear available for your choosing. Antennas, by the way, are noticeably shorter than 2-meter antennas with comparable gain.

440 MHz

The first UHF band in our upward exploration of places to operate mobile, the 440-MHz (70-cm) band covers an astounding 30 MHz-wide territory stretching from 420 to 450 MHz. To the mobile operator, the 440 to 450-MHz slice of the 440-MHz pie is generally the one of greatest interest. This is the portion of the band where you will find FM repeaters—not as many as on 2 meters, but the gap is closing.

Although the range on 440 is somewhat less than on

the lower frequency bands, linked repeaters can extend the range of your 440-MHz mobile rig far beyond the normal limits.

Get the Picture?

Just because most hams are using 440 MHz to *talk* to other hams does not mean this is *all* you can do on this versatile band. You see, it is also the lowest frequency band where you can operate fast-scan television, or ATV. Using the same type of format as the television in your living room, ATV allows you to transmit live, full-motion video—in color, if you like. Want to show your ham friends how bad the traffic on the freeway really is? Mount a camera and ATV transmitter in your car and let them see for themselves. Or use a similar setup on your boat to show your envious friends just how nice the water is as you skim across the lake. The NASCAR guys are not the only ones who can have their very own "Car-Cam"!

Of course mobile ATV can have a serious side, too. The ham with mobile ATV capability can be a real asset after a disaster strikes by transmitting live video from an affected area, providing valuable information to damage assessment officials. You may also find several ATV repeaters on the 440-MHz band in your area.

With plenty of activity, lots of repeaters, ATV and a great selection of mobile gear, it is no wonder many hams prefer 440 MHz.

1200 MHz

The last in our survey of the VHF/UHF bands, the 1200-MHz band—or as it is commonly referred to, the 23-centimeter band—stretches from 1240 to 1300 MHz. Although the band is rather large, it is only the 1270 MHz to 1295 MHz portion that will be of interest to the mobile ham. This is where you'll find FM repeaters. Granted, you will not find a tremendous number of them, but there are several, mostly in metropolitan areas. While range on the 23-centimeter band is relatively short, crossband linking of some 23-cm repeaters provides enhanced coverage.

Novices will find 23 cm appealing as it is the only VHF/UHF band besides 222 MHz where there are voice privileges. But before you buy equipment, make sure there is enough activity on this band in your area to justify your purchase!

The HF Bands

It would appear that with all the VHF/UHF bands have to offer, they must be the *only* way to go. Wrong! Don't rule out the frequencies below 30 MHz until you have read about how much fun it is to work DX—or the next state—from your vehicle.

The HF part of the spectrum consists of nine diverse bands spanning 1.8 to 29.7 MHz. They are commonly referred to as the "160 through 10-meter bands." Even though the total amount of operating spectrum offered by the HF bands is slightly less than the width of the 2-meter band alone, don't let this fool you. These "nine little bands" provide a potpourri of propagation as well as a tantalizing variety of operating opportunities to the mobile ham.

Most hams get pretty excited when enhanced propagation makes it possible for them to use their mobile rigs to talk to someone a few hundred miles away. On HF, you can do it every day. It's so easy, it's almost ho-hum.

When my family and I go on vacation, we like to "get away," sometimes traveling hundreds of miles "away." But I still like to stay informed about what is going on while I am gone. Often, as we begin our journey, I will check onto our club's repeater to chat with the locals as I drive. But, in spite of the fact that our repeater is one of the best in the area, after less than two hours on the road, its signal has faded into the noise. Even with the squelch control on my radio wide open, the occasional snippet of familiar voices soon ceases.

Am I on my own? Not a chance. I just switch off the 2-meter rig and turn on the HF rig. Connection is re-established for the remainder of the trip, no matter how far away we travel.

The Other Digital Mode

Another facet of HF you may find particularly appealing is mobile CW. Although it sounds more challenging than juggling BBs while wearing boxing gloves, many mobile operators are discovering mobile CW is both loads of fun and much easier than they thought it would be. (Most hams use an electronic keyer—instead of the old telegraph-type straight key—since the keyer requires much less effort.)

In addition, CW is inherently a more efficient mode of communication than voice. You will find your mobile CW signal will be heard farther and better than an SSB transmission using the same power level. CW is so good at getting through that most hams who like the challenge of

---

Safety First

Driving and operating can be a problem, if you let the concentration needed for driving be overshadowed by the attention paid to the radio. Mobile CW is fun—and many hams enjoy it regularly—but if you are not all that comfortable with your CW ability or drive where the road and traffic require all your concentration, then perhaps you will want to stick to voice operation.

---

running QRP (transmitter power output of 5 watts or less) use it almost exclusively. Some hams even run CW QRP when mobile. That's like bagging a grizzly bear with a pellet gun—and a lot more fun. You can leave the kilowatt amplifier at home when you go mobile HF on CW.

Has your appetite for mobile HF been whetted? Are you ready to work some DX? Maybe even give mobile CW a try? Good. Let's have a little closer look at what each of the HF bands has to offer.

10 Meters

You say you really enjoy the FM repeaters on VHF/UHF but you've grown a little tired of talking to the same people all the time? Then check out the repeaters on 10 meters. These FM machines work just like the ones you are used to—with one exception. They can frequently be heard for distances of several thousand miles. Try that on your local 2-meter repeater!

Ten meters is the only HF band where FM is allowed, but you can also use SSB or CW, in accordance with the privileges offered by your class of license. Novices find 10 meters to be especially attractive since it is the only HF band where they have voice privileges (although they do *not* have voice privileges in the subband where FM is permitted).

Although worldwide communication is often possible on 10 meters, propagation is strongly affected by the 11-year sunspot cycle—with the best and most frequent DX coming in the years at or near the cycle's peak. During the low part of the cycle the band is often—but not always—dead. Short-range communication via ground wave is possible anytime.

12 Meters

First made available to hams in 1985, the 12-meter band extends from 24.890 to 24.990 MHz. With its close proximity to the 10-meter band, 12 meters has similar propagation characteristics.

The mobile operator will find this relatively uncrowded band an ideal place for making voice or CW contacts. As you might expect, the sunspot cycle affects this band almost as much as it does 10 meters—it is great during the cycle maximum and very quiet during minimums.

15 Meters

The third largest of the nine HF bands, the 15-meter band offers the mobile ham lots of elbow room between 21.000 and 21.450 MHz. Offering worldwide propagation, it's a favorite of many hams who like to pursue DX in a more relaxed and roomier atmosphere. Daytime openings are frequent, even during periods of poor propagation. It is considered to be the best long-range DX band available to Novices, who have CW privileges there.

17 Meters

The newest of our HF bands, the 17-meter band offers 100 kHz of operating space, at 18.068 to 18.168 MHz. With worldwide communications potential, this small but still uncluttered band is becoming a favorite of many mobile hams.

20 Meters

Would you like to pass the time on your morning commute by chatting with hams all over the US? Or perhaps talk to hams in Europe or Asia as you make your afternoon drive home? Then 20 meters is the band for you.

Residing in the space between 14.000 and 14.350 MHz, 20 is the most popular of all HF bands. Under most conditions, 20 meters is open to some area of the world pretty much around the clock. Many mobile hams have used 20 meters to work 100, 200 or even 300 or more foreign countries. Especially in the evening and on weekends, however, you will be competing with well-equipped high-power stations using multielement Yagi antennas. Be prepared for a great deal of very strong QRM!

Of course you don't have to be a DXer to like 20 meters. If you frequently travel more than a few hundred miles from home, on vacation or business, you'll find the 20-meter band to be the perfect connection to the hams in your hometown.

In addition to general types of operation, you'll also find various contests and nets on 20 meters, many of them catering to the mobile operator. So whether you plan to chase DX or just work hams in the US, whether you favor CW or phone, be sure to include 20 meters in your HF mobile portfolio.

30 Meters

Weighing in with only 50 kHz of space, 30 meters is the smallest of the HF bands and runs from 10.100 to 10.150 MHz. However, don't let the small size of this band fool you. You'll find 30

Fig 2-1 – Brian Johnson, AA2OF, uses this neat installation to operate 40 and 20-meter CW. The car has 120,000 miles and 30 countries to its credit!

meters is an excellent band for mobile operation—especially if you like CW—since voice operation isn't allowed anywhere on this band. Couple that with the fact that contests aren't conducted here either, and 30 meters has the makings of a nice place to look for mobile contacts. Propagation characteristics are similar to those on the more crowded 40-meter band.

40 Meters

Beginning at 7.000 MHz and ending at 7.300 MHz, the 40-meter band is capable of providing a variety of communications opportunities to the mobile ham. Want to talk to hams in the next county? 40 will work. Want to chase rare DX? It can often be found on 40, too.

Just as with the other HF bands, propagation varies on 40 meters from day to night, as well as sometimes from day to day. That's one of the things that make this band so interesting, and fun, for the mobile operator. Of course this doesn't mean 40 is unreliable.

For 2 years, I kept a schedule on 40-meter CW with a ham 600 miles away as I drove to work each morning. Seldom were his signals less than perfect copy. Some mornings my friend Dale, NU4O, would join us as he was driving home from work, about 30 miles from me.

Although you won't find 40 meters exploding with DX the way 20 meters does, it won't disappoint you. In the early mornings and late evenings you will find 40 meters to be what hams like to refer to as "long," providing worldwide communication. When this happens, there's no telling who might show up on the band. You may find Japanese—or perhaps Russian—stations with signals sounding like locals.

When people (especially non-hams) find out I operate mobile HF, invariably they ask how far I can "talk" from my car. I always give them my pat answer: "Halfway around the world is as far as you can go. Mauritius Island, in the Indian Ocean, is just about halfway around the world from here, and I've talked to there." (And I did it on 40 meters—you can, too.)

Daytime operation is very popular, with several nets specifically catering to mobile stations. You can call in, and then ask anyone interested to go *off frequency* with you to have a chat. In the evening and at night, commercial broadcasters from Europe make this band a wild cacophony of high power carriers and noise in many areas of the country.

With a great mix of long and short range capability, it's no wonder 40 meters is the band of choice for many mobile operators—especially during the day.

80 Meters

The second largest of our HF bands, 80 meters offers 500 kHz of space, at 3.500 to 4.000 MHz. While foreign DX is possible on this band, the mobile ham will find it to be primarily a good place for ragchewing with North American stations. Conditions on 80 meters are greatly affected both by *time of day* and *time of year*. Coast to coast contacts are easily possible in the late evening hours, but ionospheric absorption so greatly attenuates signals during the day you may find it impossible to even talk farther than across town. You can also expect activity on 80 meters to dwindle somewhat during the spring and summer months—those seasons plague 80 with annoying levels of static.

During the day, 80 meters is often very quiet. Daytime propagation is under 100 miles or so. During the evening local QSOs, high power stations, traffic nets and chatting groups can make it difficult to stay in a QSO. With a well-equipped fixed station on the other end, however, you can stay in contact as you drive for several hundred miles.

Despite a few drawbacks, 80 meters is still home to many mobile hams. Various voice and CW nets, as well as a plethora of other activities, make 80 meters attractive to the ham on the go.

160 Meters

Located just above the AM broadcast band, 160 meters occupies the slot from 1.800 to 2.000 MHz. Propagation under average conditions can best be described as short range. Affected by daytime absorption and static in a similar manner as 80 meters—though more severely—160 meters is considered to be most useful during the evening hours of the winter months, when longer range propagation occurs. If you are looking for a challenge, give 160 meters a try.

So Many to Choose From . . . So Little Time

There you have it: a brief survey of the HF bands. With so many good bands to choose from, deciding which bands *not* to operate may be your hardest decision!

CHAPTER 3

Selecting Equipment

*T*he year is 2030.

You're standing in your garage alongside your 500 HP, 100 MPG, Technomobile GT-1220. In your hand you hold a Futuretronix Model ABAMATAP-300, your new HF/VHF/UHF/SHF/all-mode/ 300-watt transceiver module. About the size of an object they used to call a cigarette pack, and weighing about half a kilogram, it can easily get lost in your shirt pocket. As you admire the clean lines of the carbon-fiber encased "rig," you ponder how nice it's going to be to work DX as you make your daily, half-hour, 75-mile commute to the office on the new Superexpressway. With your enthusiasm growing, you can't wait to get that module installed in the car!

With your spoken "door open" instruction, the onboard voice recognition-command execution unit swings the driver side door open with a cheerful "please be seated" response. Flopping down into the seat, you feel the warmth from the peltier junction-backed leather seat covers as the custom interior environment-control unit warms the seat to your preprogrammed specifications. As anticipation *of your first mobile ham radio contact accelerates your pulse and respiration rates, you touch the lower left corner of the display screen of the combination onboard guidance and communications module. The screen flashes on with a, "Good evening, how can I serve you?" message.*

"Installation instructions please," you reply.

A menu scrolls onto the screen:

"Entertainment modules."

"Advanced guidance modules."

"Telecommunications modules."

"Amateur Radio modules."

You speak the command, "Amateur Radio module, please."

Across the screen comes the question, "Do you prefer on-screen or vocalized instructions?"

"Just say it to me!" you bark, surprised at your own impatience.

"Very well, please interrupt anytime you have a question," a soothing voice emanates from the 12 speaker Zobe encompass-sound system. (The voice sounds strikingly similar to that of a girl you used to date. . .) "All Amateur Radio modules constructed and marketed since 2027 are designed to plug directly into any of the four ASEDKM slots located along the bottom of the guidance and communications module display." (ASEDKM [Any Sort of Electronic Device Known to Man], a port specification standard devised and implemented at the ISO 35000 convention back in '25) "Do you have a module that you wish to install?"

"Yes I do," you reply in a slightly calmer voice, wondering to yourself why you feel remorse for screaming at a computer.

"Please read aloud the electronic device identification number," the VIAB (voice-in-a-box) requests.

"One, nine, five, five, seven, one, six, four, zero," you read.

"Please verify that this is the Futuretronix ABAMATAP-300 module."

"It certainly is," you reply.

"Please insert the module into any available slot, making sure that the EDIN hologram is facing up."

Obediently, you do what the VIAB has instructed.

"Please stand by," says the VIAB, *"initialization is being performed."* Less than two seconds later, *"Initialization complete, please stand by for additional information and options selection."*

Has Fact Overtaken Fiction?

This fictionalized account of a self-configuring rig in the year 2030 may have already been overtaken by technology in the mid-1990s. Long distance telephone and communication companies have had systems that automatically select the route of your telephone call or data message, depending on traffic load, conditions, available satellites and maintenance tests. The Institute of Electronic and Electrical Engineers (IEEE) in the May 1995 issue of *IEEE Communications Magazine* describes the new generation of radios, consisting of an RF front end followed by an A/D (analog to digital) converter. From this point on, software controls the radio. Wideband, narrow band, AM, CW, SSB or picture modes can all be selected by software. Change bands, change modes, change speeds—all by software commands.

Voice control can be bought today at your local computer store. Several systems allow you to move, copy or delete files by voice commands. Several of them also allow you to "pick up" and answer your telephone (built into your computer) with brief voice commands.

Still think this account of ham radio is far out? Maybe it is closer than you think!—*Paul Danzer, N1II*

Your mind wanders as you wait, and eventually you realize you are thumping your thumb on the steering wheel. You wonder if maybe it wouldn't be a good idea to cut out the coffee.

Moments later, you're jolted back to reality by the VIAB: "Your Technomobile GT-1220 has the most advanced electronics system known to the twenty-first century. The electro-isolated passenger compartment roof not only protects occupants from the elements, it also serves the additional function of an auto-tuned, multiplexed antenna for all radio frequency transmitting and or receiving modules you may wish to install. In addition, the thought interpretation and execution capability of the GT-1220 control system is available for use with the Amateur Radio module—at your option of course. Since the Amateur Radio module you have installed is the 300-watt version, you must select antenna time sharing priority for it and the previously installed sat-phone module. What percentage of time do you wish to make available to the Amateur Radio module?"

Your first inclination is to select 100 percent for the Amateur Radio module, but since the sat-phone is crucial to your occupation as an artificial intelligence researcher, you acquiesce.

"50 percent," you reply, hoping no one will ever phone just as you snag a rare DX station.

"50 percent confirmed," replies the VIAB. "Do you wish to allow control of the Amateur Radio module by the thought interpretation and execution capability of the GT-1220 control center?"

"Yes I do," you reply.

"Thank you," the VIAB returns, "option selection is now complete. Enjoy your new Amateur Radio and please be careful when operating as you drive. You're only human, you know."

"Yeah, right," you shout, more than a little incensed. Then you muse, thinking aloud, "That voice sure does sound familiar. . . could it be? Nah, couldn't be. Could it. . .?" Glancing at the clock, you see that it's just after 7:00 PM. "Boy, I wonder if there's any DX on 20-meter CW right now," you think to yourself. Instantly, the car is filled with the melodious sound of several CW signals, all of them rare DX. As you listen to the music pouring from those 12 speakers, you suddenly realize that one of the signals is a ZF2 calling "CQ USA," and you need him for DXCC. No sooner does the thought enter your mind than the ZF2 is instantly the only signal you hear; all others have been automatically filtered out by the digital signal selector feature of the ABAMATAP-300.

"Wow, this is great," you shout. You then think, "call the ZF2 at 40 words per minute and give him a 599." The ABAMATAP-300 does as you've requested, and you've snagged another rare one. . . .

Sound good? Of course it does! If we just knew that the year 2030 was going to be that good for ham radio! What with cigarette pack-sized all-mode, all-band radios, controlled by our thoughts and feeding built-in stealth mobile antennas—why we would all head down to the local cryogenics lab and give them instructions to thaw us out in 2030!

Meanwhile, Back in the Present . . .

But wait! Things aren't so bad right now. Sure, the current crop of Technomobiles doesn't exactly offer lots of good mounting options for our radios. To make matters worse, once we've somehow figured out how to squeeze a radio or two into our vehicles, we sometimes find that the automakers haven't paid a lot of attention to designing cars that are good electronic neighbors with our equipment. And with many new autos costing more than some of us paid for our homes just a few years ago, it's pretty hard to get up the nerve to set a mag-mount antenna on the roof, much less to even think about *drilling holes*!

Don't let all that discourage you. In spite of those obstacles, when it comes to choosing our mobile radio equipment we hams have never had it so good. While it's true that we can't go out and buy a Futuretronix ABAMATAP-300 (yet!), the current generation of compact, feature-packed mobile ham gear is vastly superior to anything available even just a few years ago. In fact, if this book were being written in the not-so-distant past, this chapter would have consisted largely of (gasp!) schematics and construction information showing you how to *build* your own mobile gear. Sure, there's something very satisfying, and perhaps a bit romantic, about operating equipment that you've built from scratch; a good number of hams still do it. But it's unlikely today that you'll find many hams who mourn the passing of the days when you *had* to build your own mobile gear!

So there's nothing left to do besides run out, buy a radio and install it. Right?

Wrong! If you're the impulsive type, you can easily find yourself in trouble if you leap before your look. Once you've made up your mind to buy a new piece of gear, it can be easy to get seduced by the pretty pictures and the slick sales talk. You'll be happier in the long run if you use those sales techniques as a jumping-off point for your research. Yes, research. As once-exotic features become standard, once-reasonable costs can soon get out of hand. Unless you figure out what features you want in a mobile rig, you could be stuck with far more—or far less—radio than will serve your particular needs.

Before you pull out that credit card, check the Product Reviews. (The ones in *QST* are based on the results of sophisticated, independent lab tests.) Read all the manufacturers' literature you can get your hands on. Ask around—perhaps a member of your club just bought one of the rigs you're considering. See how she likes it before you invest in the same model. Aside from the monetary considerations, your choice of mobile ham gear can directly affect your ability to operate your vehicle safely. The next section goes into detail on these and other factors that should go into any purchasing decision.

Playing It Safe

If you are going to operate your mobile radio while you are driving (and if you aren't, you won't be mobile, you'll be *portable*—or you'll be chauffeured) then the attention required is stolen away from the attention you devote to your driving. It doesn't take a genius to figure out that a very difficult-to-operate radio for mobile use can be a prescription for disaster. If you operate such an unruly animal at your home station, at most it might cause occasional frustration or maybe a missed contact every now and then. Run a mobile rig with that same sort of disposition and you could wind up missing much more. Always remember that the most important thing when it comes to mobile operation is safety, safety, safety!

While we're on the subject of safety, it's worth noting that in recent years most of the radio manufacturers have become much more conscious of the need to address ergonomics—especially with mobile equipment. Ergonomics is just a fancy way of saying that the machine (or in our case, the radio) should be designed to be easy and convenient for the human to operate.

That sounds simple enough, but it's not always so easy

Portable, Mobile and Signing your Call

The FCC requires us to identify ourselves by transmitting our call signs at the end of each contact, and every 10 minutes during the contact. On phone modes this must be done in English, and on CW at a speed of 20 wpm or less. This is all that's required! However, as a courtesy to your fellow hams, it is customary to provide a little more information.

Generally, you would say you are mobile if you are using a station capable of operating while in motion. Car, truck, bicycle or on foot—it makes no difference. A station that is not at your license location could be called portable—especially if you will remain at this location for an extended period. Many hams add the call area to their identification. For example, while on vacation in Arizona, I like to sign *N1II portable 7*. While driving around my home town in Connecticut, I usually use *N1II mobile 1*.

Al Brogdon, K3KMO, operates mobile CW from both his car and his motorcycle. He says many mobiles on CW use */M* and others use */M3* (telling the other station they are mobile in the third call area). Still others operate mobile CW without any addition to their call sign.

The FCC Rule Book, published by the ARRL, contains the full set of FCC requirements. No matter how you classify your station, and no matter how you sign your call, you must meet the minimum FCC requirements. Anything more is up to you.—*Paul Danzer, N1II*

to accomplish. For instance, back in the days when a 2-meter mobile rig had a front panel with only three knobs—the on/off-volume control, squelch control and channel selector—it was pretty hard to get confused while operating the rig. About the only thing that would invoke frustration on the part of the operator was if the channel indicator light burned out! Contrast that with today's multi-featured 2-meter rigs with upwards of a dozen front panel controls, some with dual or even triple functions, and it's easy to see why ease of operation is often one of the most important considerations when shopping for a mobile rig.

Keep in mind too when you are shopping for a mobile rig that while it's always good to ask for your friends' opinions and recommendations about the radios they've used, it's possible that your tastes may be quite different. That's why it's imperative to become thoroughly familiar with any rig you may be considering for mobile use. The best way to do that, of course, is to actually sit down and operate the rig, simulating as closely as possible the conditions under which you would be operating it if it were installed in your own vehicle. Does it seem intuitive to operate? For instance, if it's a VHF or UHF FM rig, are the more commonly operated controls located on the microphone? Does a simple command such as switching from a memory channel to the VFO require you to look at the rig (and away from the road)? If it does, then you may want to consider a different radio. While you may find that there's no rig made that you can operate without ever taking your eyes off the road, your first priority should (must!) be to spend a minimum amount of time looking at the radio and a maximum amount of time with your eyes on the road.

Another essential consideration, unless you plan to never operate mobile at night, is whether or not the controls you will most often use are illuminated. As you shop around you may find rigs that have addressed this consideration rather well; others haven't addressed it at all.

Keep in mind that having to turn on the dome light in order to see how to access a particular function of your rig doesn't make for safe motoring.

An additional way to determine how convenient a rig will be for you to operate mobile is to examine the operator's manual carefully. Does a simple operation such as turning on CTCSS encode have such a steep learning curve that you must read the instructions several times before you can successfully enable that option? If so, you may find that operating that rig while mobile isn't as easy as you'd like. Check it out!

Granted, with so many features being packed into such small packages, some functions must be activated by multiple presses of a particular control—or sometimes by the simultaneous use of more than one control. Obviously, this is a price we pay for wanting—and getting—radios that do so much while occupying so little space.

But what if you don't need all that complexity in a mobile rig? Well, you may want to consider one of the "barebones" rigs that some manufacturers are now producing. Those rigs concentrate on the basic operating necessities. This not only results in a rig that's usually got a clean, uncluttered control panel and a relatively smooth learning curve, it will mean you'll probably save money on the purchase price. No sense paying for features in a radio you'll never use.

Regardless of whether you select a radio with Cadillac-like sophistication or Volkswagen-style simplicity, be sure the features you are likely to use most often don't require you to divert your attention from your driving. Always remember that using a radio should *never* compromise the safe operation of your vehicle.

Okay, we've spent some very important time driving home the need for safety considerations. Now let's move on to some other criteria you may want to use in making your selection of a rig for mobile use.

50 MHz and Up

The most popular form of mobile operation is on VHF/UHF. If this is your first shot at mobile hamming, VHF/

UHF will probably be your initial choice as well. So let's have a look at some of the features you may want to consider when shopping for mobile VHF/UHF gear.

Be Remote Without Being Aloof

There's no question that the bane of present day installation of mobile ham gear is the diminished space that's available in the modern automobile. What's surprising is that even though VHF/UHF rigs have steadily shrunk in size, we find fewer and fewer available places to mount them in our vehicles! What to do? Go remote, of course.

And you know what? That's not a new idea at all. You see, back in the earliest days of mobile ham stations, many hams—and Amateur Radio manufacturers—found that by providing a separate control box for a mobile rig, the bulky, heat-producing main chassis could be mounted in a remote location, most often the trunk. Needless to say, that made for much less crowding in the passenger compartment, which translated into happy hams and even happier family members! Nor was that lesson lost on the manufacturers of commercial communications equipment. As a Field Service Station technician in the early 1970s for a major

Fig 3-1—This Kenwood TS-60S is a 6-meter all-mode transceiver. It provides full SSB, AM and CW capabilities for local contacts as well as DX when the band is open. The FM mode allows you to work through both local repeaters and simplex. All this comes in a 6.4-pound mobile package measuring approximately 2.5×7×9 inches. The only thing not light is its power output— slightly under 100 watts for most modes (30 watts on AM).

manufacturer of business-band radio equipment, practically every mobile installation I dealt with consisted of a remote-mounted main chassis with an under-dash control head.

In the world of Amateur Radio, however, an interesting change had begun to take place by the 1960s. With the arrival of the transistor, it suddenly became fashionable, as well as practical, to make mobile gear self-contained and designed for under-dash mounting. Not that the radios were necessarily all that small; by today's standards, they weren't. It just happened that as Amateur Radio gear began to get more compact automobile interiors were also becoming more expansive. A convenient combination of events, no?

But as we all know, half of that happy equation didn't last. With the gas crises of the 1970s and government-mandated fuel economy standards, we suddenly found the interior space of our vehicles shrinking faster than a cheap sweater in a red-hot clothes dryer. Even with the current generation of mobile rigs that are smaller than ever before, the arrival of unsupportive plastic dashboards, full-length center consoles and downsized interiors have made it difficult to find a place to mount ham gear in many vehicles.

Fortunately, the Amateur Radio equipment manufacturers have come to our rescue with remote-mountable radios for mobile use. As the name implies, on a rig with remote-mountable capability, the main radio chassis can be mounted in a remote location, perhaps under a seat or even in the trunk. Then the detachable control panel, usually rather small and easily mounted, can be positioned in a convenient location that's easily visible and accessible by the operator. A cable (usually part of a remote-mounting kit available at additional cost from the manufacturer) then connects the main chassis of the radio to the detached control panel. This sort of installation has obvious advantages. No longer do you have to worry that your under-dash mounted rig may someday end up on the floor, with a large chunk of the plastic dashboard still attached. Nor do you have to worry about cutting up a portion of your car's interior to make a place to install your radio, thus dooming the resale value when you go to trade.

Salesman: "What's the big hole in the dash for?"

Ham: "Hole? What hole? Oh, that's an optional air conditioning duct I ordered on the car when it was new. It's really a rare and valuable option, you know."

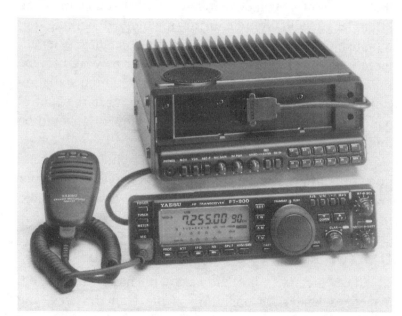

Fig 3-2—For the would-be mobile HF operator with limited vehicle space, the Yaesu FT-900 is a dream come true. Taking its cue from the popularity of remote-mountable VHF/UHF rigs, the FT-900 is the first HF rig to be manufactured with a removable control panel. With the main chassis tucked away in a safe location, the control panel can be mounted in an easily accessible spot, making almost any vehicle suitable for an HF installation. The FT-900AT is an optional version of the FT-900, and comes with a built-in antenna tuner.

Salesman: "Sure buddy, and I'll just bet that Mario Andretti is your uncle too!"

Seriously though, in some vehicles you may find that a remotable rig is the only practical way you can even operate mobile.

There can also be a fringe benefit to having a remotable rig that you might not have considered. You see, if you install the remote control panel so that it is easily removed when you leave the vehicle, and then complement it with an easily removed antenna, you have a very effective theft deterrent. If you do a great deal of traveling, or if you often leave your vehicle in an unattended parking lot, remote mounting of your radio gear is a good idea. After all, there's not much worse than having *your* new rig become *someone else's* new rig. Concealing your radio installation can be accomplished much more quickly when the control panel can be easily removed from the view of a would-be thief.

As you can see, the advantages to remotable gear are numerous. So if you've come to the conclusion that your vehicles' interior isn't very hospitable toward a radio installation, perhaps a remotable rig is the answer.

It's Two Rigs—Two Rigs—Two Rigs in One!

Two meters is hot! In fact, with more than 10,000 2-meter repeaters in operation in the United States, you could say that 2 meters cooks. But it's not the only burner on the stove. More and more, the 440-MHz band is gaining enthusiastic advocates.

Once considered by some to be just a hangout for egghead types with rooms full of rack-mounted equipment that had parts with names like "nuvistor" or "varactor tripler," 440 MHz is now called home by more than 5000 FM repeaters. So, with a total of more than 15,000 repeaters

VHF/UHF Rig Features

The ARRL monthly journal, *QST*, often carries comparative product reviews. These let you compare features and performance of similar rigs from all the major manufacturers. But what happens if there has not been a review of the rig you are considering? Then you make your own comparison table.

Along the top, list the model number of each rig you want to compare. On the side, list the features available. Where do you get the list of features? From the manufacturers' advertisements. Each month *QST* will carry four to six full page advertisements, usually on the inside and back covers. These advertisements list the features of the rigs. Combine the lists, make a check mark if a rig has that feature, and you can now make a good comparison. The list below is a good starting point. Each month the manufacturers seem to add features, so don't take this list as the last word!

Features

- Number of memory channels
- CTCSS encode/decode
- CTCSS frequency search
- Built-in VOX
- Power out and power selection
- Location of DTMF tone pad
- Controls on the microphone
- Dual band
- Cross-band repeat function
- Wideband receive coverage
- Aircraft band receive
- Can be modified for MARS and CAP
- DTMF paging
- DTMF autodialer
- Hands-free accessories
- All keys and controls illuminated for night-time use
- Display dimmer
- Large, visible display of all functions
- Scan modes
- Can be tuned and controlled without the instruction book.
- Price, including tax and shipping if applicable

shared between 2 meters and 440 MHz, no matter where you might travel throughout the country, you're likely to find either of these two most popular bands bustling with activity. So how do you choose which band is best for you? The good news is *you don't have to choose.* You can operate both bands from one rig by using a "dual-bander."

As the name suggests, a dual-band radio is one that has the ability to operate on two different bands. While various combinations of band coverage are offered (and some rigs even offer coverage of more than two bands), the most common dual-band offering is the combination of 2 meters and 440 MHz. With a dual-band radio, a whole new world of mobile operation opens up to you. Not only do you have access to nearly twice as many repeaters as you would with just a single band rig, you also gain the capability to operate *duplex.*

Here's how it works. Normally, when we talk on a radio, regardless of whether we are using FM, SSB or even CW, we

Fig 3-3—It's two rigs in one! This ICOM IC-2340 has independent controls for 2 meters and 440. A total of 110 memories gives you a large choice of preset frequencies. Almost everything, including telephone dialing numbers, is programmable.

can only hear the person we are talking to when we stop talking and listen. That's because we talk and listen, or technically, we transmit and receive, on the same frequency. That's what is known as *simplex* operation. On the other hand, with a dual-band radio, we are no longer limited by that restriction. Because a "dual-bander" can simultaneously receive on one band while transmitting on the other, you are able to hear the person you are talking to even as you speak. That's duplex operation, and it allows a more natural style of conversation. It's as if you are using a telephone instead of a radio. That's a particularly nice feature when you are making an autopatch through a cross-band repeater. The person you are talking to on the autopatch doesn't have to wait for you to unkey your radio before he or she can reply or interrupt.

Of course duplex operation isn't limited to autopatch operation. It doesn't even require you to use a repeater. But regardless of whether you use a repeater or talk direct, the ham on the other end will also have to be using a rig with duplex capability.

Have Your Own Repeater

The ability of your mobile rig to transmit on one band while simultaneously receiving on another makes available another very attractive option known as *cross-band repeat*. Available on practically all dual-band radios, it adds a slight twist to the normal form of duplex operation. Although the rig is still receiving on one band and transmitting on another, instead of the transmitted audio coming from the microphone, it is now coming from the receiver. To put it another way, whatever is being received on one band is simultaneously being retransmitted on the other. Think of it. You have your very own mobile repeater! What's that good for? Glad you asked.

Consider the case of a ham friend of mine. He's an outdoors type of guy who does a lot of hunting and fishing.

During deer season, he frequents a rather remote area that's ruggedly landscaped with deep valleys and jutting hills. It's the kind of terrain that renders his HT virtually useless, even to work the local repeater. Not wanting to risk having a situation arise where he might need emergency assistance, but be unable to reach someone on the air, he hit on a reassuring solution. He simply installed a dual-band rig with cross-band repeat capability in his truck. Now when he ventures deep into the wilderness on his hunting trips, he parks his truck on a hill overlooking the area where he's going to hunt and then sets up his mobile rig to cross-band repeat. With 50 watts and a good antenna, his mobile rig has no problem getting a full-quieting signal to and from the local repeater. And since he's never more than a couple of miles from his truck, his dual-band HT works his mobile rig with very little effort. Obviously, he now hunts with a great deal more peace of mind.

Of course that's but one example of how handy your cross-band repeat mobile rig can be. Some hams use their cross-band repeating radios along with their dual-band HTs to stay in touch with the gang on the local repeater while they browse deep inside the shopping malls—where it's sometimes hard to hit any repeater, unless of course it's sitting in the park-

---

A Little Caution

If you do use your rig as a repeater, keep in mind it was probably designed for a low duty cycle. Many rigs will overheat when they are forced to transmit for extended periods, as may happen if you allow your rig to re-transmit the output of the local repeater. You might be able to select a low power output setting or only pick this repeater function during periods of low use of the local repeater. Remember, you are responsible for the output of your rig. Always make sure you can shut it off, if it is transmitting something it shouldn't.

---

ing lot. Other hams dispense with the need to have a second VHF/UHF rig and antenna for their homes by leaving their mobile rigs in cross-band repeat mode and using only an HT from the house. Now that's a way of using cross-band repeat that becomes even more attractive when the mobile rig is equipped for *remote control*. That simply means that you can use your HT, or any other rig with a DTMF pad, to remotely control some of the functions of your mobile rig.

Say for instance you have your mobile rig set for cross-band repeat and you are using it to retransmit your HT's 440-MHz signal to 146.52 MHz, where you are chatting with a friend. As he travels farther away, he starts to lose your signal and he asks you to move to the local repeater. If your mobile rig has remote control capability, all it takes is just a few key presses and, presto, you've switched over to the repeater. You are now continuing your conversation without having missed a word. And you didn't have to run out to your car to do it. That's why it's called remote control.

By the way, handy as cross-band repeat and remote control are, keep in mind that when you use those functions, there are several FCC rules, and sometimes other restrictions, that must be observed. We'll look at those in detail in a later chapter.

All Good Things. . .

Hopefully this isn't beginning to sound like a sales pitch for all the new mobile rigs. With so many good features available—dual bands, remote mounts, repeat, extended receive, remote control, dual in-band-receive and well, you get the idea—it's sometimes hard to determine just exactly what capabilities you'll need, and want, in your mobile installation. The preceding "wish list" of desirable features, from a mobile operator's standpoint, should get you started on your search for the right VHF/UHF mobile rig. As always, thor-

Fig 3-4—If you are planning to take a long trip, and want to store the repeater frequencies you might want to select, the 50 memories in this Yaesu FT-2200 will probably be enough. Standard offsets (plus or minus 600 kHz on 2 meters) and "oddball" splits can be stored. The extended frequency coverage on receive includes automatic selection of AM when an aircraft band frequency is chosen.

oughly gather your information and consider it carefully before you buy. And by all means, ask lots of questions. After all, it's your money. And that's the bottom line.

We Now QSY to HF

Were all you prospective HF mobile operators thinking you were going to be left out? Not a chance. Let's take a look now at some possible criteria to use when selecting an HF rig for mobile operation.

It's a Nice Rig, But Where am I Going to Put It?

What's one of the most difficult aspects of performing an HF mobile installation? If you guessed that it's finding a satisfactory location to mount the rig, you are right. Think it's hard nowadays to find a place in your car for your

2-meter rig? Pity the poor HF operator who's dealing with a rig that's considerably larger.

Fortunately, the situation really isn't as bad as it seems. Just as with VHF/UHF gear, manufacturers are beginning to make greater strides in the direction of more easily mounted HF mobile rigs.

Shrink Wrap

One very logical approach is to make the rig physically smaller. As electronic technology has advanced, giving us surface mount and other space saving components, it has made

Fig 3-5—With the 555 Scout, the folks at Ten-Tec have taken a unique approach to mobile HF rigs. The Scout dispenses with all the bells and whistles common to most HF rigs, concentrating on basic SSB and CW coverage of the 160 to 10-meter bands with 50 watts of RF output. CW operators will appreciate the Scout's built-in keyer and full break-in capability, both standard equipment. The absence of non-essential frills, as well as an unusual band-switching arrangement that uses individual plug-in band modules, help to keep the cost of this small (2.5×7.25×9.75 inches) rig down to about half that of most HF mobile rigs.

possible the packaging of more and more features into an ever smaller box. This has resulted in a new generation of HF rigs that, while still somewhat larger than their VHF/UHF siblings, are much more space conscious than ever before.

At least one manufacturer has taken downsizing of HF mobile gear a step farther by leaving off the "bells and whistles" and offering only the most essential functions necessary for an HF mobile rig. The result is a package only

Picking an HF Rig

The suggestions made earlier in the chapter, in the sidebar on picking a VHF/UHF rig, hold, for the most part, for HF rigs as well. The starting point is the *QST* reviews, and the ending point is a table similar to the one given for the VHF/UHF rig.

Features

- Size compatible with your car
- Single or multi-band
- Built-in antenna tuner
- Ease of control and number of controls (since you will be looking at the road, and not the rig!)
- Can you lock the controls?
- 12-volt input requirements
- Enough audio power to overcome road noise
- Internal speaker location and external speaker jack
- Effective noise blanker
- Mobile accessories
- Price, including tax and shipping if applicable

One other note on picking an HF mobile rig—pick one with enough flexibility for your use. Single-band rigs are fine, but recognize their limitations. A 10-meter-only rig during the low part of the sunspot cycle will not provide many QSOs, nor will a 20-meter SSB-only rig do very well if you plan to operate only on Sunday afternoons, when the band is full of kilowatt stations!

a smidgen larger than a 2-meter rig, yet offering complete coverage of the HF bands.

And it gets better! If you can live without all the unnecessary frills, as well as making do with 5 watts or less of RF output, you may want to consider one of the QRP rigs being offered by several manufacturers. Depending on your operating needs, your choice of QRP rigs ranges from the ultra-small single-band CW-only rigs to multi-band rigs offering the AM, CW, SSB and FM modes as well as many of the features found in their larger brothers.

But what if you don't want a stripped-down rig with limited features? And QRP isn't the kind of challenge you're looking for when operating mobile; but you simply don't have enough room to mount a conventional HF rig in your vehicle? Well. . . you may want to consider an HF rig that has remote control capability. As with the VHF/UHF rigs that offer the same option, you can place the main radio chassis in a convenient spot that's out of the way, then mount the removable control panel in a location that's easily and *safely* accessed.

As you can see, finding a place in your vehicle to mount an HF rig has never been easier, *if* you choose the right equipment.

Which Bell? Which Whistle?

Okay, now that we've solved the "I can't find a rig that will fit in my car" problem, what else is there to look for when choosing an HF mobile rig? Well. . . that depends on what sort of operating you plan to do. But regardless of your planned mode of operation, there are a few attributes that you'll always find useful.

Give Me No Static, Please

One is a good *noise blanker*. Designed to suppress the various forms of electrical interference that we commonly

Fig 3-6—Can't make up your mind? This ICOM IC-706 covers HF plus 6 and 2 meters, all modes—in a very compact package.

refer to as static, a noise suppressor can be a life saver for the mobile operator. Although you might be lucky enough to have a vehicle that generates little or no static of its own (though it's unlikely!), the vehicles you share the road with might not be so well mannered.

Keep in mind also that automobiles aren't the only source of electrical noise. High tension power transmission lines are another hideous source of noise, and they seem to always run alongside the roads we travel most. Since your vertically polarized mobile antenna is a very efficient gatherer of all that electrical garbage, an *effective* noise blanker can be a really valuable asset to your mobile rig. Note too the emphasis on the term *effective*. I've owned some rigs whose noise blanker circuits seemed to have been installed strictly for cosmetic reasons and were next to useless for suppressing noise. In fact, on one rig I owned (the names are being withheld here to protect the guilty), the blanker circuit was so poorly designed that not only was the performance lackluster, when it was activated it allowed so much IF blow-by past the CW filter that it was like switching the filter out of the circuit.

Point: If you are shopping for an HF mobile rig, a real-world evaluation of the noise blanker performance is a wise move. Remember that just because it says "NB" on the panel doesn't necessarily make it so.

Built-In Antenna Tuners

Once available only as outboard accessories that were often bigger than the rigs they were used with, antenna tuners are now available as a built-in option for many mobile-sized rigs.

Do you need one for *your* mobile rig? That depends on how you plan to operate.

If you intend to tune your antenna and rig to 3950 kHz and leave them there, an antenna tuner will probably be a waste of money, especially if the rig you decide to use only provides a tuner as an option.

On the other hand, if your operating plans include frequent excursions between band edges (especially on 40 and 80 meters where most mobile antennas have very limited bandwidth), and you don't want to have to stop and adjust your antenna each time you QSY, then you will probably benefit from a built-in tuner.

Beware too that a tuner won't allow you to use an antenna that's not made for the band you're operating. Even if the tuner should achieve a match, the performance of the antenna on the wrong band will be dismal at best!

Lock It Up

Another feature that's really desirable in a mobile rig is a VFO lock to disable the tuning knob. Once enabled, this control prevents accidental changes in operating frequency.

Want to become notoriously unpopular with your

fellow hams? Then let your rig do a vibration-induced QSY boogie all over the band as you transmit.

Dahdidahdit Dahdahdidah

Are you considering joining the rapidly growing ranks of mobile CW operators? If so, you may be thinking that your needs in a rig are much different than for other modes. Fortunately, they're not. As a rule, the preceding list of features will be an asset to the CW operator as well. There are, however, a few things you will want to look for if you are planning to become a "brass pounder" on wheels.

Keyers

It seems that hardly anyone pounds brass anymore, especially when mobile. Instead, we relegate all that "dit-dah" work to electronic keyers. And even though it sometimes appears that CW capability has been added to some HF rigs as an afterthought, most rigs offer an internal keyer as either standard or optional equipment. Since the last thing most mobile installations need is added cabling and another box to mount, it makes sense to use a rig with a

---

To CW or Not to CW

Mobile CW is fun. It provides the advantages of CW—better copy, usually less interference from high power stations and even a more relaxed pace for the QSO. It does take concentration, especially when looking for a station to contact, adjusting the rig to reduce interference or trying to copy a weak station.

Driving also takes concentration, both in city traffic and at highway speeds. If you are not sure of your CW ability or driving conditions are poor for any reason, you should be prepared to abandon the idea of CW and either stick with phone or shut down completely. Safety First!

---

built-in keyer, right? Of course! While you're at it, don't forget to check out how the keyer speed is controlled. Unless you will be setting the speed once and then forgetting it thereafter, which is *very* unlikely, it will pay you great dividends to see how easily the speed can be set while you are mobile.

On some rigs, you may find this control to be very touchy to adjust, making it extremely difficult to get the keyer speed you want. Other rigs have very easily adjusted controls, and some even give a display panel read-out of keyer speed. Which type of control would you prefer to use?

Filters

Anyone who's done much operating in the crowded CW bands will testify to the value of a good CW filter. But to the mobile CW operator, a filter isn't a luxury, it's a necessity. And not just because of the need to eliminate adjacent frequency interference, or *QRM*. A narrow-bandwidth CW filter is also very effective in reducing noise. Here's why.

The noise you hear on the HF bands is wide-band in nature, regardless of whether it is atmospheric, ignition or power-line generated. Consequently, the wider the bandwidth of the filter in your rig, the more noise you receive. Install a narrow filter, and the signal-to-noise ratio of your rig improves dramatically. That means that CW signals that may have been barely discernible with a wide filter will rise above all the racket when you switch in a narrow one.

Reduced noise and QRM when you operate mobile CW can greatly reduce operator fatigue. And fatigue is something you neither want or need when you're driving. So get a mobile rig with a CW filter—one with a 500 Hz bandwidth is a good compromise between ease of tuning and effective performance. Your ears will thank you.

Whew! We've looked at a lot of good features to consider when choosing a rig for mobile operation, but before we leave the topic of equipment selection, let's look briefly at choosing a vehicle.

Deals On Wheels?

What does choosing a vehicle have to do with selecting equipment for mobile operation? Plenty! Have you heard the story about the guy who was out several hundred dollars to replace the electronic control unit in his vehicle after it died from exposure to RF from his HF rig? Well, that and many other horror stories of incompatibility between ham gear and automotive electronic systems have been widely circulated in recent years. And while it can sometimes be hard to separate fact from fiction in some of these tales, one thing is for sure—there's no guarantee that your ham gear is going to coexist peacefully with all the electronic widgets found in the modern automobile. Gone are the days when you could solve the worst case of interference in a mobile installation with resistor spark plugs, suppression wires and perhaps a few bypass capacitors. It's a new day—with plenty of new challenges.

So what should you do? The subject of RFI, both *to* and *from* radio equipment, will be dealt with in-depth in a later chapter. But if you are planning to trade vehicles soon, or you are planning a mobile installation in a vehicle that you've never operated radio gear from before, you may want to do a little detective work first.

One form of RFI that's easy to check for is that which originates from the vehicle electronics and interferes with your radio. Since several of the electronic devices in a modern auto, from the AM/FM radio/clock to the engine control computer, are generators of radio frequency energy, it's no surprise that much of the QRM we deal with in a mobile installation comes from within. Question is, will that RFI hinder your ability to operate? The only way to know for sure is to survey the candidate vehicle.

This is especially true if you plan to operate an HT on the VHF or UHF bands with only the attached flexible antenna.

Case in point: Something in my Dodge Daytona Shelby renders my HT nearly useless on the 2-meter band by generating a very strong, raspy signal at 1-MHz increments, starting at 144 MHz. That wouldn't be so bad except that it sometimes also generates a full-scale signal on the same frequency as the local 2-meter repeater, usually right when I'm in the middle of a QSO. Fortunately, an outside antenna cures the problem.

If you plan to operate the VHF or UHF bands, and especially if you are going the HT/rubber duck route, try using that setup to scan the entire band of interest as you observe for interfering signals that are originating from the vehicle. What you find may affect your mode of operation, choice of vehicle, or both.

What about cases where our radio equipment adversely affects the electronic systems in our vehicles? With the potential for expensive damage, coupled with the fact that not a lot is known on the subject, this is a pretty hot topic. For now let's just say that if you are not sure what effect your radio installation will have on a particular vehicle, do some research, starting with the vehicle manufacturer. Then carefully read the chapters on installing equipment and dealing with RFI problems.

New or Used? That is the Question

Although our primary focus in this chapter has been on the selection of *new* mobile equipment, many cost-conscious hams will probably consider the purchase of a pre-owned radio. If you choose to go that route, all the previous guidelines will still apply. However, if the radio you are considering is no longer being produced, you might have difficulty getting information about its particular features. Usually, a call to the manufacturer will provide you with a general overview of the radio's capabilities. (Many factory technicians will also make you aware if the model you are considering has exhibited a record of poor reliability.)

Of course, if you are doing your shopping at a hamfest flea market (which presents its own hazards!), time won't allow you the luxury of doing a lot of prepurchase research. In that case you may want to prepare yourself by having with you a copy of a buyer's guide for used equipment— such as the *ARRL Radio Buyer's Sourcebook*. A compilation of *QST* Product Reviews, these books are an invaluable aid to the buyer of used equipment—detailing the specifications of each radio as well as the reviewer's impressions of its operation.

CHAPTER 4

Installation

roper installation is the key to a safe and successful mobile setup, guaranteeing many years of enjoyable operation. In this chapter, we'll look at the various steps involved in performing mobile installations—both temporary and permanent. But before we drill that first hole, or make that first connection, let's consider a subject important to *all* mobile installations.

Location *Is* Everything!

No matter *how* you plan to operate mobile, your rig must be located in such a way as to not compromise your ability to drive safely! Remember, operating your radio is a *secondary* operation, *driving* is your *primary* responsibility. To this end, both *where* and *how* you mount your radio equipment will be the most important aspects of your mobile installation. With this in mind, it's a good idea to precede any radio installation with a thorough survey of the passenger compartment.

Does the prospective mounting location offer you a

clear, unobstructed view of the display and controls? I once saw an HF rig installed vertically against the dash of a car with the display facing the roof! It was necessary for the driver to lean forward until his head almost touched the dash in order to see what frequency the rig was tuned to!

Not only should you be able to see the controls, it's imperative you also be able to easily reach them. Leaning over in the seat as you stretch across the car to reach your radio is a quick way to make unintended lane changes—or worse. If you cannot find a place to mount your rig where you can easily operate it as you drive, find another vehicle, another rig—or limit your operating to times when you are a passenger.

As you scout for a suitable location for your rig, be especially observant for any potential interference with your vehicle's safety equipment. With air bags popping up in all passenger cars built in the US since 1993, you must be absolutely certain you don't mount equipment where it can hinder or prevent the airbag's deployment. If you are unsure if your vehicle is equipped with an airbag, or of its location, contact your dealer for assistance. You might also wish to refer to the April 1993 *QST* article "Don't Get Blown Away by Your Mobile Rig," by Brian Battles, WS1O, for some helpful information on the topic of airbags and mobile radio installations.

Of course, airbags aren't the only safety feature found in the interior of an automobile. Padded dashes used to be offered only as optional equipment, but they've been around so long now we pretty much take their "face-saving" qualities for granted. Over the years, auto safety engineers have taken great strides in designing dashboards and other interior components with impact-absorbing materials. Be careful you don't negate the effectiveness of those materials by mounting a rig in a location where you or a passenger might strike it during an accident.

Newton's First. . .

Do you like science experiments? I have in mind one you might enjoy. It's fun, interesting, educational and pretty easy to perform. All you'll need is an assistant, your mobile rig and a 12-story high building. Got all those things gathered together? Then here's what you do.

Instruct your assistant (your spouse will probably be willing to help) to take your radio and go up to the roof of the building. While your assistant is doing that, you should position yourself on the sidewalk in front of the building. When your assistant gets to the roof, he or she should, on your signal, toss the radio off the building toward you.

You may wish to take cover immediately after you give the signal!

Neat experiment, huh?

Okay, this experiment wouldn't be very much fun for *you* or *your radio*. But consider this. When the radio reaches you, it will be traveling at about the same rate of speed as it would be if you were driving down the highway with it in your car. If your vehicle should become involved in a collision, you might find your nice, friendly radio coming at you (or a rear-seat passenger) with a vicious force equal to that of a sledgehammer pounding spikes into seasoned railroad ties! But *not* if you have taken the time to secure it properly. Sure it's tempting, especially in temporary installations, to forego the proper mounting of a rig. Resist that temptation as though your life depends on it. And be sure the mount really *is* secure. A 12-pound HF rig can exert *hundreds of pounds* of force upon its mounting hardware in the milliseconds of extreme deceleration caused by a collision. Don't let your rig become a deadly missile! Make it secure.

To Do. . . Or Not To Do?

It's been said the best way to get a job done is to have someone else do it. Does this axiom apply to the job of installing a mobile rig? It depends on you. Are you confident in your ability to route cables, mount equipment, perhaps even to drill holes in your automobile? Some people are understandably reluctant to attempt that sort of task, especially with a new vehicle. But you shouldn't let the prospect of making a costly mistake while performing a radio installation deter you from operating mobile. Instead, you might want to consider having your radio equipment *professionally installed.*

Admittedly, you won't find a listing for "Amateur Radio Installers" in the yellow pages of your telephone directory. In fact, few, if any, of the dealers who sell amateur equipment even offer installation service. (The long-held reputation of hams as do-it-yourselfers *must* be a factor.) Who are you gonna call? The people who install and service the radios in police cars, fire trucks and other emergency vehicles, that's who.

Actually, any shop installing commercial or *land mobile* two-way radio equipment should be capable of installing amateur gear. After all, most of them have done hundreds of mobile installations, some of them possibly in vehicles just like yours—and that's valuable experience. Not only that, if interference problems crop up, most shops are willing, for a fee, to seek out and correct them. If you check around, you might even find a shop in your area employing one or more of your ham friends. For the ham who doesn't want to perform his or her own installation, these shops are a good alternative.

If you decide to have your equipment professionally installed, here are some suggested guidelines:

- Be sure the shop you select understands what type of equipment they are dealing with. (Some shops are willing to do Amateur Radio installations but won't install CBs or consumer electronics.)

- Discuss beforehand where you wish to have everything mounted.
- Establish who will be liable for any possible damage to the radio or vehicle.
- If on-the-air testing of the completed installation is planned, make certain a licensed ham is present as control operator. It is illegal for an unlicensed person to transmit on the amateur bands for *any* reason.

You *Can* Do It!

Perhaps the easiest, and arguably the most common way to operate mobile is with a hand-held transceiver, or *HT*. Powered by its own battery and equipped with a rubber duck antenna, installation of an HT is as simple as climbing in your car and driving away. Besides easy installation, the HT route has other advantages. Security is one. You never have to leave your rig unattended when you leave your car. And, if you frequently drive different vehicles, it's not necessary for you to install a radio in each one. (A real advantage for the ham who often uses a rental car.) With the immense popularity of HTs (as a visit to any hamfest will confirm), it's easy to see why they are so commonly heard transmitting from the asphalt arteries that criss-cross our continent. Unfortunately, some of the features that make them so convenient can also pose a problem for the mobile operator. How many times have you heard a QSO similar to this?

K9MUT from K4IQR. I'm sorry Jake, but I didn't get all that last transmission. You were in and out of the repeater. Could you repeat?

Sure, Bob. I said I'm on the HT. . . (hiss, crackle) . . . only a rubber duck and . . . (more hiss, more crackle) . . . battery may be going. . . (silence).

Is there hope for K9MUT? It's hard to say, but his radio installation is unquestionably a prime candidate for some improvement.

If you plan to use an HT while you are mobile, carrying a spare battery can prevent you from having to sing the "There Goes the Battery" blues.

A spare battery can rescue you in a pinch, but you might want to go one better and connect your HT directly to your car's electrical system. Before you do, determine if your radio allows a direct connection to 13.8 volts (13.8 is the nominal voltage rating of an automotive electrical system, but a maximum system voltage of 14.5 is common). The owner's manual of the radio should provide you with the needed information. By the way, don't assume just because the wall charger for your HT is rated at 13.8 volts, you will be able to connect the HT to 13.8 volts in your car. Not only can excess voltage damage your radio, an overcharged NiCd battery can explode! (Please don't ask how I know.)

Choose carefully your method of tapping into your car's electrical system. The fuse panel is a popular place to extract the needed "juice," but making good connections can sometimes present quite a challenge. (It also requires you to locate a good grounding point nearby, which can be a *real* exercise in frustration.) Perhaps you'll find it more convenient to make your connection at the cigarette lighter jack, using a plug made for this purpose. That will not only spare you from doing an impromptu "Pretzel Man" impression as you grope under the dash for the fuse panel, it also makes moving your HT "connection" from one vehicle to another as easy as pulling a plug. Having your HT "wired" to the vehicle electrical system can ensure your HT battery doesn't go flat at the worst possible moment—and it will often enable your HT to produce full output power—

A Different Sort of Installation Problem

Bicycle Mobile! My first mobile installation was on an old Schwinn bicycle, with heavy rubber tires (the only kind available) and a wire basket. Batteries and a 2-meter rig filled the basket between the handlebars, and the whip antenna extended from the top of the rig. Bicycles—and bicycle mobile—have come a long way. The *Bicycle Mobile Hams of America* is an organization for hams who are also bicyclists. They have an on-the-air net and a newsletter. For more information and a sample newsletter, contact them at PO Box 4009-L, Boulder CO 80306, telephone 303-494-6559.

It's not quite turnpike speed operation, but you can get the fun of two hobbies at once.—*N1II*

something many HTs *don't* do when operated from a standard battery.

No matter where you make your connection, be sure the power leads coming to your radio are properly fused—don't depend on the existing automotive fuse for protection. Why not? It's common for a cigarette lighter circuit to be equipped with a 20 to 30-amp fuse, and that's sufficient current to cook even the toughest HT. Also, be *absolutely* certain that you check your connections for the correct polarity *before* plugging in your HT. Once while driving home from a hamfest, I loaned my HT's dc cord to a friend so he could check out the HT he had just purchased (the battery was not charged). Imagine our astonishment when my friend's new toy spewed forth a cloud of green smoke as soon as he plugged in the cord. Even though his HT had the same type and size power jack as my HT, the polarity was *reversed*. (In case you're wondering, yes the cord *was* protected—with a 5-amp fuse. The fast-blow fuse was protected by some faster-blowing components in the HT. Murphy strikes!)

Reaching Out

Having thwarted the "dead battery" syndrome, it's possible you'll still find the range of your HT to be inadequate at times. This is when the addition of an outside antenna can help it behave like one of the "big boys." While some hams use a permanently mounted antenna to boost the range of their HTs, a magnetic mount (or *mag-mount*) antenna is more in keeping with the "temporary" flavor of HT mobile operation. We'll be taking an in-depth look at the selection and installation of antennas in the next chapter—but if you opt to use an external antenna with your HT, choose one using coax cable with a *stranded* center conductor. The constant flexing of the coax where it connects to your HT will inevitably break a solid center conductor, putting you off the air without warning.

Know what HT stands for? Handy Talky. Having power and antenna cables dangling from your HT as you attempt to operate it and drive at the same time can seem about as handy as wrestling a rattlesnake. How do you tame the savage beast? The best way is with the addition of a *remote speaker/mike*. With one of these clever devices, you'll be able to leave your HT stowed safely away in its mounting bracket while you are engaged in a QSO.

You *are* planning to provide a mounting bracket for your HT aren't you? You haven't heard? Unsecured HTs are notorious for slinking around the inside of a car like hungry rats, hiding beneath the seat, munching on old French fries and sometimes dancing between the driver's feet. Seriously, you should *always* provide some means of securing your HT, no matter how you plan to use it. Several companies manufacture suitable mobile brackets for HTs, or perhaps you'll want to fabricate one yourself. In his March 1994 *QST* article, "An Over-the-Dash H-T Mount,"

Herbert Leyson, AA7XP, describes a simple mount he devised to secure his HT while mobile.

Going for the Long Haul

Although the "HT on wheels" approach is ideal for the car-hopping ham, you may find the day-to-day rigors of *your* style of mobile operation beg for the use of a conventional mobile rig—mounted in a slightly more permanent fashion.

If you are one of those fortunate hams who own a vehicle with generous interior proportions, an *under-dash* installation will most likely be the easiest and most sensible way to go. Tucked safely away under the dash, your rig will not only be less likely to be in harm's way, it also will be much easier to see and operate. As an added bonus, if you've equipped your rig with the proper mounting hardware and cable connectors, it will be a simple chore to remove it from the vehicle when you leave, providing increased security—or making dual-usage of the rig possible.

If you are planning an under-dash installation, be aware some automobiles have dashboards that are *very* unsupportive. This can be especially true if you are installing an HF rig. Not only does the dash have to support the weight of the rig, it also has to contend with the leverage produced

Fig 4-1—Richard Scott, K4JQA, built this wood base for his rig. The rig is held in place by a *bungie cord.* **The raised front gives him a good viewing angle of the rig's front panel.** *(Photo courtesy of K4JQA)*

by the rearward mounting location of most rigs in their brackets. To circumvent possible damage to the dash, choose a metal point of attachment, if at all possible. If you cannot find any metal along the bottom of the dash (not uncommon), try to locate an area having some type of reinforcement above it—frequently, this will be where a brace is attached. Should that fail, it may be possible to place a narrow strip of metal on the upper side of the lip of the dashboard, providing the needed reinforcement. On larger rigs it's advisable to provide a rear hanger or bracket for additional support.

Holy Dashboards, Batman—He's Got a Drill!

If the automotive engineers have been kind, there's a chance you'll be able to use existing holes—and if they've been really kind, existing screws or bolts—to attach a bracket for your rig. Lacking such good fortune, some soul-searching may be in order as you struggle to decide if going mobile is really worth giving your car the Swiss Cheese treatment.

Why should you drill holes?

Simply put, in many cases it's the *only* way to adequately secure a rig.

Won't drilling holes in your car destroy its resale value?

Maybe not. Sometimes, especially with under-dash installations, you can drill *invisible* holes. An invisible hole is simply a hole that won't be visible (at least not to a prospective buyer of your vehicle) when you have removed your radio equipment. If the invisible hole is a little more visible than you'd like it to be, a bolt, screw or push-in plug of the right type and finish can often leave things looking factory original.

Isn't it risky to drill holes in today's cars?

It certainly can be. The two-way radio shop where I

worked had to foot the bill for a very expensive repair job after one of their technicians drilled through the floorboard *and* into the transmission case of a new car. If you find it necessary to drill holes *anywhere* in your vehicle, follow some simple precautions.

- Check thoroughly *behind* where you plan to drill to be sure there is nothing that can be damaged by the drill.
- Use a center punch to mark the hole—this prevents the bit from "walking" and doing damage.
- Use a stop collar to prevent the bit from penetrating any deeper than necessary.
- Before drilling through carpet, make a short slit with a sharp knife. This will prevent the bit from grabbing the yarn and unraveling it.
- Be especially careful when drilling holes in the trunk—gas tanks aren't always located where you expect them to be.
- Pick your drill carefully. Use either a 3-wire drill with a 3-wire extension cord or a 2-wire double-insulated drill.

Consoles and. . . Cubbyholes?

While floor shifts and consoles stretching from the dash to the back seat give a car that nice, sporty look, they can sure give fits to the ham who's bent on installing a mobile rig. With an under-dash installation pretty much out of the question in this type of auto, what's a ham to do?

Obviously, if you already have the car, and you are shopping for a rig, then one of the best options is to choose a rig with a removable control panel. Not only will you be able to place the control panel just about *anywhere* that suits you, the companion mounting bracket can usually be affixed with double-sided tape, or something similar, eliminating the necessity for drilling holes. Of course it will

Use Your Existing Car Sound System For Better Mobile Hand-Held Audio

How many times have you heard or said on the local repeater, "Say again—I'm in a noisy area"? There's only one answer: *too* many! Fixes for this problem usually employ outboard audio amplifiers and external speakers that require 12-V connections and lots of cabling. I amplify my radio through my car's existing cassette player and sound system via a CD-player-to-cassette adapter. This device consists of an audio coupler in a cassette shell that plugs into a portable CD player via a cable and 1/8-inch stereo plug. (I use the CD adapter that came with my CD player. Radio Shack sells a similar product.) For ham radio use, I convert the adapter to monaural operation, which allows my car sound system's balance and fader controls to route the radio audio to just one speaker or all four. You can do this conversion three ways. The first and easiest way is to buy a 1/8-inch stereo-to-mono adapter (RS 274-368). The second method involves removing some of the plastic between the tip and ring of the CD-adapter plug, soldering the tip and ring together and removing any excess solder with a file. Although this successfully shorts the right and left channels together for radio use, it also ruins the adapter for its intended purpose. (If you don't already have a portable CD player, you may get one someday, so consider this option carefully before doing it.)

I prefer a third method that includes RF-feedback suppression and takes a little more work. Cut the cable between the adapter and its plug and install the circuitry shown in the photo. (This is a good time to figure out the optimum length for the adapter cable and size it accordingly.) Install the filter as close as feasible to the cassette adapter shell; this maximizes its effectiveness in suppressing RF feedback in transmit. The entire circuit can be constructed on a small circuit board and housed in a 35-mm film canister. S1 allows you to switch between stereo and mono without using a plug/jack converter—a plus if you lose small parts like I do. Keep the switch from being bumped accidentally by mounting it on the board facing the canister cap.

This system worked well in two of three cars; the third car's system required 0.01-μF capacitors across its speakers to eliminate RF feedback. When I use my converted adapter, keeping the hand-held/mobile volume high and the stereo volume low seems to minimize noise. Adjusting the sound system's tone control for more bass than treble also seems to help.—*TAB Brown, NM3E, Sinking Spring, Pennsylvania*

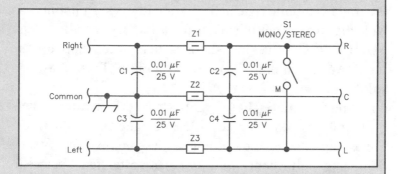

TAB Brown installs this circuit in his CD-to-cassette adapter cable to add stereo/mono switching and suppress RF feedback. Z1-Z3 are ferrite beads or low-value inductors; a few turns of enameled wire, $^{1}/_{4}$ inch or so in diameter, will also work. C1-C4 are Radio Shack 272-131 discs; S1 is an RS 275-065. The filtering components (Cs and Zs) may not be necessary at transmitter powers under 25 watts.

still be necessary to find a good, out-of-the-way location to mount the main chassis of the radio—the trunk is usually the best choice. Avoid mounting *any* radio equipment—transceivers, amplifiers, et al—under the seat. Poor ventilation there can cause your equipment to experience potentially damaging high temperatures, not to mention that some power-operated seats can crush an improperly located under-the-seat rig. In addition, some automakers place electronic control modules under the seat—all good reasons *not* to mount equipment there.

The new, remoteable rigs are nice, but if you don't own or plan to purchase one, you may choose to do an *in-dash* installation of a conventional rig. There's no question mounting a rig in the dash provides for a very unobtrusive and aesthetically pleasing installation. Properly installed, the in-dash mobile rig can blend in with its surroundings so well that it is often mistaken for a factory-installed item. Nevertheless, some hams are reluctant to chop up the interior of an automobile to install a radio. There may be a solution.

Many vehicles have small storage areas or pockets built into the dash or console. Designed to hold tissue boxes, tapes, CDs or other items, these compartments (one automaker calls them *cubby holes*) are often ideal places to mount a VHF/UHF rig. They are usually both tall and wide enough to allow installation of a rig, but may be a bit shallow, not providing enough depth for the radio to fit flush with the front. Even if there *is* enough depth, you still must provide an exit point for power and antenna cables (be sure your rig's heat sink or cooling fan isn't obstructed). You can often overcome those obstacles by carefully cutting off the vertical portion of the back wall of the compartment, allowing the rig to extend out the rear (Fig 4-2). Don't discard the piece you cut out. It can be glued back in place when you remove the rig, maintaining the resale value of your vehicle.

Or better yet, cut the piece out of someone else's vehicle. When I traded vehicles a few years ago, I too was faced with the necessity of doing some cutting of the dash in my newly acquired wheels. Once I determined what pieces would have to be cut, a quick trip to the local auto salvage yard yielded me the same parts from a similar vehicle—and for a reasonable price too. (All mobile hams should be on a first-name basis with the proprietor of the local auto-recycling emporium.) Back at home, I removed the original parts from my car and stored them safely away for later installation at trading time. The salvage yard parts were then modified and installed in place of the originals. I'm happier, and so is the car.

Fig 4-2—The rear wall of this storage compartment has been carefully removed, allowing installation of a small mobile rig. The cooling fins and rear panel connectors of the rig extend through the new opening. Make sure you have under-dash clearance for the protruding connectors before you cut the rear cover off. Save the cut-off section, and you will be able to re-glue it together when trade-in time comes for the car.

Even if your car doesn't offer you the convenience of a cubby hole for mounting a rig, there are other possibilities. Sometimes you can remove a map light, or perhaps an ashtray, to make space for a rig. In one of my vehicles, the area

Fig 4-3—N1II removed the unused ashtray from his minivan, and with a little judicious filing replaced it with his Kenwood rig. A block of wood, cut to fit the cup holder, has a microphone clip screwed to the top to hold the mike. *The ARRL Repeater Directory* rests on the utility shelf above the broadcast radio. *(Photo courtesy of N1II)*

once occupied by the ashtray now houses an on-board performance timer.

Regardless of how you approach an in-dash installation, be sure to take time and plan *before* you begin. *Accurate* measurements are a *must*. The rig obviously has to have *enough* room to fit where you want it, *and* you want to be absolutely certain you don't remove more material than required, leaving your rig framed by nothing but open space. As you measure, be sure there is enough clearance *behind* the dash for your rig *and* any required connections. Remember too that while it may have taken quite a while for you to install your rig, it will take a thief much less time to remove it. You may want to protect your rig by using an auto security system or by fabricating a cover plate to hide it when you leave the vehicle.

If neither an in-dash or under-dash installation is feasible, you may have to resort to mounting your rig on the floor (common for larger HF rigs). If this is the case, you'll need to either purchase or fabricate a suitable mount for your rig. In my February 1993 *QST* article "*You* Can Operate HF Mobile!" I described one method of mounting an HF rig on the floor of a car. As with any installation, make sure the rig and its mount are suitably secured. Screws or bolts through the floorboard are the best plan here. . . but watch out for those transmissions! (Automotive, not radio.) Remember to slit the carpet before drilling, as previously mentioned, and the holes won't be visible when the equipment and screws are removed later. Also, it's a good idea to place an anti-seize compound on any bolt or screw that extends outside of the passenger compartment, just to ensure you can remove it when you are ready.

Getting Wired

A neatly installed rig is a beauty to behold, but it's only so much window dressing until the necessary power and antenna cables are routed and connected. Although this phase of the installation will leave you more intimately acquainted with the inner being of your automobile than you may have expected (or desired), don't worry. By following a few simple ground rules, your success is all but guaranteed.

You *Can* Get There From Here

The ideal installation is one leaving no wires or cables exposed where they can be damaged or present a trip hazard to passengers. If you'll be running any cables front-to-rear through the passenger compartment, it's preferable to place them *under* the carpet, along the bottom of the transmission tunnel. This location has some advantages.

It's an area of low wear and impact, so there's less chance of damage to the cable.

It helps to provide greater separation from vehicle wiring, which is usually routed along the doors under the *sill plates* (those are the plastic or metal strips running along the top of the rocker panels to cover the edge of the carpet).

Above all, be sure to avoid running power or antenna cables near any electronic control modules. You'll most often find them located under the passenger side of the dash, usually behind the kick panel. Unfortunately, they can also be lots of other places. If you're not sure of their location in your particular vehicle, contact your dealer's service department *before* you make your cable runs. Your dealer may also be able to provide you with valuable technical information from the manufacturer pertaining to mobile radio installations. Or you may wish to obtain a copy of the General Motors *Mobile Radio Installations Guidelines* by sending your request, along with an SASE, to the EMC Department, 40-EMC, General Motors Proving Ground, Milford, MI 48380.

Getting Connected

Remember when you installed an eight-track tape player in your second-hand 1966 Dodge Dart? Did you strip off about an inch of insulation from the negative power lead and slip it between the dash and the tape player mounting bracket just before you tightened the last bolt, clamping the wire firmly in place? Perhaps you then snaked the positive power lead along the back side of the dash, across the steering column and over to the fuse block. Once there, you may have pried one end of a glass fuse out of its socket (the one labeled "radio" was a favorite), fanned out the strands of the lead, inserted them into the vacant fuse clip and then pressed the fuse gently back into place. Presto! In only a matter of minutes, you were ready to groove to the stereophonic sounds of your favorite group as you cruised the local drive-in.

While wiring equipment that way may have worked for a tape player—in 1970—today's mobile rigs require a much different approach. To begin with, each piece of equipment should be individually and directly connected to the vehicle battery. Connecting to the battery instead of to existing vehicle wiring has several advantages. First, it can help to eliminate reception of electrical system noise. Also, the battery tends to dampen voltage spikes and surges (after all, it's really just a gigantic capacitor). You've probably heard that your car's starter induces a very unhealthy voltage spike into the electrical system when it is disengaged. What you may not have known is that when a malfunctioning alternator that's stuck in full-charge mode loses its connection to the battery (and thus its load), it can pump more than 75 volts into the electrical system of the vehicle. So go straight to the battery, with both the negative and positive leads. This means you'll be running *two* leads to the battery for each unit to be powered. (A transceiver and outboard amplifier will require a total of four power leads, for example.)

It might appear to be easier to simply connect the ground lead to the chassis of the vehicle instead of to the battery, but it's not advisable.

"Okay, but can't I just run one set of power leads from the battery, then feed all my equipment from them?"

You could. Here is why you shouldn't: VOLTAGE DROP. The power leads for your equipment are sized according to the amount of current they need to carry. Try to power several devices from a lead that isn't large enough to carry the combined current, and they'll *all* end up starved for voltage. Modern microprocessor controlled rigs can do *very strange* things when the supply voltage drops. If you aren't using the factory-supplied power cable for your equipment—or if you have to increase the length of the leads, be sure you use wire of sufficient size.

Breaching the Wall

Does the firewall appear to be an impenetrable barrier as we journey toward the battery? Take a second look. On most

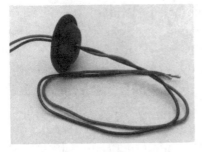

Fig 4-4—You can use a rubber plug, with a hole drilled in its center, as a firewall feedthrough grommet for your cables.

vehicles you'll find lots of holes in the firewall. It's simply a matter of picking one to use. If it's available, choose an unused hole that has been fitted with a rubber or plastic plug. Those are the holes for equipment not installed on your particular vehicle (for example, many automatic transmission-equipped vehicles have an unused hole in the firewall where the clutch cable or linkage would have passed through). If you put one of these idle holes to work, don't throw away the plug. By drilling a small hole in it, you can use it as a *grommet* to protect your wiring and to reseal the hole, as shown in Fig 4-4.

If you cannot locate an unused hole, it's usually possible to persuade existing hoses or cables to share their grommets with your cables. Perhaps the most obvious (and visible) choice will be where the factory wiring harness, a large bundle of perhaps as many as 50 wires, passes through the firewall. Unfortunately, it's also the least desirable choice, for a couple of reasons. Many of the wires in those bundles are connected to electronic devices, posing the possibility for harmful interference. (If there are two bundles passing through the firewall on your vehicle, the one on the passenger side is especially a must to avoid. It usually connects to the engine control module.) In addition, the grommet is often *molded* onto the harness, leaving very little room for extra wires to pass.

Holes for hood release cables, clutch cables or vacuum hoses all offer a better point of entry for your power cables. Of course, if all else fails, you can make your own hole in the firewall. If you must do that, avoid possible damage by accurately locating the place you plan to drill. Make the hole no larger than necessary and be sure to seal the hole with the correct grommet, to prevent the entry of fumes and noise into the interior of the vehicle.

Once they are through the firewall, route your cables carefully along the driver's side of the engine compartment, avoiding exhaust system components, belts, pulleys and anything else that moves. Use wire ties to keep your cables in place while maintaining as much distance from vehicle wiring as possible.

Getting Disconnected?

Radio equipment manufacturers stipulate that to prevent possible damage to your rig, you should always turn it off before cranking the car engine. Since you are connecting directly to the battery, you can't rely on the ignition switch to cut the power to your rig when the starter is engaged. To eliminate having to always remember to turn the rig off, you may want to install a cut-out relay in the power lead to disconnect the rig automatically when the starter is engaged. Check the August 1994 *QST* "Hints and Kinks," page 67, to see how John Conklin, WDØO, built a cut-out circuit for his mobile rig. Be sure to choose a relay having contacts rated to handle your rig's input current.

The best way to make connections to the battery is with a suitable terminal, attached to the existing cable clamp hardware (see Fig 4-5). Although you've probably seen it done, it is *not* advisable to remove the cable clamps, place the ends of the power leads into the clamps and then replace the clamps on the battery posts. Here's why. The posts on a top-post battery are *tapered,* and so are the holes in the cable clamps. When properly mated, the post and clamp form a tight fit with a large contact area—necessary for the large amount

Fig 4-5—Here's a good example of how to make your connections at the battery. Sandwich the ring terminals between two flat washers and use an additional nut to hold them in place. Coat the surfaces with plenty of conductive anti-seize compound.

of current they must carry. When you place a wire between the post and clamp, the misalignment you cause not only reduces the mating surface area and makes the clamp difficult to keep tight, it also allows acidic fumes to corrode the post-to-clamp junction. Eventually, the rapidly deteriorating connection will leave you with a car that won't start.

Be sure to observe polarity when making your connections at the battery. It's also a good idea to leave all equipment **disconnected** from the power leads until those connections are made (pull the fuses if necessary). This will not only save you much grief if you inadvertently hook things up wrong (double check your connections), it will also help to prevent *unwanted* sparks near the battery.

There's a Bomb Under the Hood?

Your car's battery looks harmless sitting there in its black plastic wrapper, doesn't it? Don't let that fool you. Treat it with the same respect as you would a live hand grenade. Here's why.

As a result of the occurring chemical reaction, lead-acid storage batteries give off hydrogen gas when they are being charged or discharged. Normally, the hydrogen is harmlessly dissipated into the atmosphere. But, given the right conditions, a spark can ignite the gas and cause the battery to explode—hurling battery fragments hundreds of feet and showering the area with sulfuric acid. There have been many instances where lost eyesight and bodily disfigurement resulted from battery explosions.

Keep in mind too that batteries don't have to explode to do lots of damage. Even though they supply relatively low voltages, they can supply tremendous currents that rival most arc-welding machines. A misplaced tool or bare wire bridging the terminals of a battery can be heated to a searing white hot in milliseconds. Take these basic precautions when working around batteries.

- Wear eye protection, or better yet, a face shield.
- Don't place tools or other conductive objects on top of a battery.
- Avoid creating sparks when making connections to a battery.
- Keep spectators at a safe distance.

Keeping Your Car Smoke-Free

Fuses are cheap life insurance for your rig and your car. Make sure you properly fuse *both* power leads. Think a fuse in the negative power lead is a waste? It really isn't. If you examine your vehicle's battery cables, you'll find the large negative cable is connected directly to the engine block. That's because it must handle the return current from the starter, which can easily exceed 100 amps. If the cable should lose contact while the starter is engaged, your rig's case and negative power lead may attempt to complete the circuit. Just one additional 10 cent fuse could save your rig.

Make sure the fuses are located as close as possible to the battery connections. This prevents the possibility of fire in the event a power cable becomes shorted to the chassis. The power cables for most mobile rigs now come with the fuses at the battery end; if yours didn't, you'll need to relocate the holders. But don't let fuseholders rest on top of the battery. Fumes from the battery will corrode the terminals, making disassembly of the fuse and its holder difficult or impossible.

Stay Connected

Vibration, moisture and temperature extremes can team up to deliver a knock-out punch to your mobile rig's power and antenna connections. Over time, connections that were once sound can deteriorate, causing electrical interference, equipment malfunctions and possibly even a fire. A few simple steps can help keep your connections secure.

Make sure all mechanical connections are tight. Use lockwashers where needed and periodically check nuts, bolts and cable connectors for loosening. If you find it necessary to splice wires, it's best to solder them and cover the splice with heat-shrink tubing. If you choose to use crimp-on terminals, be sure to use the proper crimping tool to install them. Where possible, take the time to also solder the wire to the terminal to ensure a good electrical connection. Crimp terminals that cannot be soldered, as well as any nut, bolt, screw or washer used for power or

The Traveling Ham's Survival Kit

Odds are you can motor around your local neighborhood for the next 50 years and never experience a problem with your mobile rig or antenna. But hit the road for places far away and more than likely some part of your mobile installation will hold your QSOs hostage with an intermittent connection, loose mounting hardware or perhaps just a blown fuse. If you have packed your Traveling Ham's Survival Kit, you can quickly pay the ransom and be back on the air. If you haven't, you may find some towns don't have a Radio Shack store.

The list here is far from all-inclusive; you'll want to modify it to suit your own installation, travel agenda and determination to get back on the air. The idea is to take along the most basic items you might need, and hope Murphy's law does not apply—after all, a screwdriver is a screwdriver, right?

- A VOM and miscellaneous tool to fit battery connections, rig mounts and antenna hardware. Include Allen wrenches and metric sizes.
- Electrical tape—don't leave home without it! The same for the manual for your rig, unless you have memorized all 2461 programming commands for your dual bander.
- Spare fuses—both for your rig and for the auto. If the air conditioner fuse blows while you are operating, your ham gear will be blamed.
- The *ARRL Repeater Directory* or the *ARRL Repeater Atlas*. These come in very handy in a strange area if you need help finding a motel or restaurant. The *Repeater Directory* also comes in electronic form, perfect for laptop computers.

antenna connections should have a conductive anti-seize compound such as *Kopr-Shield* applied to prevent corrosion. You may also want to add some further insurance against corrosion by using stainless steel or brass hardware.

Sound Solutions

You've seen it on television. The private detective places a drinking glass against a wall, puts his ear to the glass and eavesdrops on the conversation in the next room. If you've ever tried it yourself, you would probably agree the sound could best be described as being somewhat hollow, having a far-away quality.

Unfortunately, the audio from many mobile installations has very similar characteristics. Granted, tiny speakers in tiny rigs are obviously a *big* part of the problem. Further compounding the problem is that most rigs have only two possible locations for the built-in speaker—on the top or bottom cover of the rig. Consequently, the sound coming from your under-dash rig either gets directed toward the floor, where the carpet absorbs it very nicely, or it gets bounced off the bottom of the dash and subsequently ricochets haphazardly about the car, creating a not-so-pleasant reverb effect. The in-dash rig fares even more poorly, its audio sounding like it's coming from the car in front of you.

Unless your mobile operation is casual and very infrequent, you'll probably want to provide an outboard speaker for your rig. Many companies manufacture remote speakers, some with built-in amplifiers, suitable for mobile use. You can also find used or surplus commercial mobile two-way radio speakers at many hamfests. These speakers are ruggedly designed and provide good quality sound—and are usually very reasonably priced. Be sure the speaker you choose is the correct impedance for your rig.

When installing an outboard speaker, keep in mind the best results will be obtained by mounting the speaker as close as possible to your ears, aimed so the sound is directed toward you—especially important if you frequently travel with non-ham passengers who haven't developed a love for the musical quality of ham radio. In fact you may want to

Fig 4-6—If you can't hear them you can't work them. Mark Goff, WA4JSN, mounted two speakers to direct the audio in his car toward his ears. He managed to squeeze them in without blocking any controls.

mount the speaker right over your head. In some vehicles, there is enough space between the interior headliner and the roof to install a speaker. Normally, all that's necessary is to remove the trim along the driver's side edge of the headliner and gently pull the headliner down far enough to insert your speaker. Route the speaker leads behind the windshield pillar trim and then to the radio. Unless you've got lots of room, you'll probably have to use a small speaker with no enclosure. Be sure to secure the speaker so it doesn't wander about. (The overhead speaker approach really works great for mobile CW operators. My XYL enjoys our vacation travels much more now, since she doesn't have to listen to 45+ WPM CW while we are on the road.)

Overhead Consoles

Many new mini-vans and other autos now come with an overhead console. Some, such as the series from Chrysler Corporation, include a sunglass holder and a bin for a garage door opener transmitter. If you are not using one of these, the holders can be used to contain a small speaker.

Even without using one of these bins, the overhead console makes a good mounting place for an external speaker, but be careful to select a position away from your head. If the speaker is mounted on the outside surface of the console, you should be able to swing your head toward the passenger side without encountering the protruding speaker.

Stealth Speakers

Before you dash out to purchase an outboard speaker for your mobile installation, you may want to investigate the possibility of using components of your existing vehicle sound system. Several approaches are practical. Some hams provide an interface between their rig and the car radio, allowing them to use the entire audio system, as TAB Brown, NM3E, described in a *QST* "Hints and Kinks" column (see the sidebar, "Use Your Existing Car Sound System for Better Mobile Audio," earlier in this chapter). Alternately, you may wish to feed your rig's audio to only the speaker(s), but if you do that, provide some means of isolating your rig from your car radio, to prevent possible damage.

If you are hesitant to risk making connections to your car's sound system, you may be able to install your own speaker in an existing but unused factory speaker location. Since most automobiles have many sound system options available, there is almost always a place where you can install an additional speaker. For example, many autos with factory-installed stereo sound systems will have an unused speaker location, complete with grill, in the center of the dash. (This is an excellent location. The windshield focuses the sound back toward the passenger compartment.) A trip to your dealer or the local auto salvage yard may net a factory correct-fitting speaker.

Don't Lose Your Key

Anyone who has operated mobile CW can tell you it opens up a whole new world of ham radio excitement. For long-lasting enjoyment of this unique form of mobile operation, it's essential you mount your key or paddles properly. Unlike a microphone, the key must remain stationary during operation. Some hams use a leg clamp mount, which works fine if you aren't getting in and out of

Fig 4-7—The author has used this arrangement in several cars equipped with bucket seats. The platform is made from 2×6-inch wood stock. Not visible are two feet, also made from 2×6-inch stock, extending from the platform to the floor. A short length of metal strapping is attached between the legs and screwed to the floor with a sheet-metal screw. One of the rubber feet on the paddle base has had its retaining screw replaced with a longer screw. This screw passes though the platform and holds the paddles in place.

your vehicle frequently. If the seat in your vehicle has a fold-down arm rest, it may be possible for you to mount your keying device there. This can provide the added convenience of being able to fold the key up and out of the way when it is not in use.

Fig 4-7 shows a mount I have used successfully in three different vehicles equipped with bucket seats. The platform keeps the paddles firmly in place and its added length in front of the paddles provides a convenient forearm rest—a must to prevent operator fatigue during extended periods of high-speed CW operation.

As you can see, setting up a ham station on wheels isn't a plug-n-play proposition. So take your time, plan carefully, proceed even more carefully, and your mobile installation will be a source of pride and operating enjoyment for a long time to come.

Stay tuned as we move to the next chapter and address one of the most talked-about topics in ham radio: Antennas.

CHAPTER 5

Antennas

 hat we have here is a failure to communicate." —Strother Martin in *Cool Hand Luke*

Buy the best, most expensive, gizmo-of-the-week radio equipment you can find, install it in your car, hook it to a poor antenna, and it's likely you too will experience a failure to communicate! The antenna provides the launching pad for your signal, so the better the antenna and its installation, the greater the reward—more effective and enjoyable communications.

VHF/UHF Antennas

As you peruse the manufacturers' catalogs or the ads in Amateur Radio magazines, you'll see a mesmerizing variety of VHF/UHF mobile antennas. Which one should you choose? Your choice will be affected by many factors, such as cost, method of mounting, performance, appearance, even the height of your garage door. Don't laugh—one ham I know ripped his newly installed dual-band antenna completely off the roof of his truck the first time he drove into his garage.

Dual Antennas vs Dual-Band Antennas

If you will be using a dual-band rig, you'll need an antenna for both bands. Should you use separate antennas for each band, or one for both? If your rig has a diplexer built in, this decision is made for you. A diplexer combines the signals for both bands into a common feed line. If your rig doesn't have a diplexer, cost may be a factor. Using separate antennas not only saves you the expense of buying a diplexer, the combined cost of two antennas may be less than the cost of some dual-band antennas—but you will have to mount two separate antennas.

On the other hand, if you prefer to mount only one antenna, you'll need a dual-band antenna. Of course, everyone knows dual-band antennas are a compromise, offering lackluster performance. Well . . . if you compare specifications, you may be impressed with the gain figures posted by some dual-band antennas. See the sidebar for a discussion of advertised antenna gain. By using a long radiating element (some 2-meter/440-MHz antennas approach 5 feet in length) and appropriate loading coils, some dual-band antennas boast greater gain than that of many single band antennas.

Be aware however, that in addition to being rather tall, many of these high-performance dual-band antennas are very rigidly constructed, making them extremely unforgiving of collisions with low-hanging fixed objects. One antenna manufacturer's representative told me the rigid construction and lack of a spring base helps to keep the antenna vertical at highway speeds, thus enhancing antenna performance.

If you choose one of these large antennas, you may either want to select one with a foldover mast, or use a foldover mount to allow the antenna to be lowered out of harm's way.

The Gain Game

As you shop for a VHF/UHF antenna, you'll often see references made to the term *gain*. What is antenna gain and why is it important?

Gain is a relative measurement of an antenna's performance, stated in decibels or *dB*, as compared to a standard, or *reference* antenna. If the reference antenna is a half-wave dipole, the antenna gain will be listed as *dBd*, or gain over a dipole (a half-wave dipole and a quarter-wave vertical have equal gain). On the other hand, if the reference antenna is an isotropic radiator (a theoretical antenna that radiates equally in all directions), the gain will be expressed as *dBi*—for gain over an isotropic radiator. If the antenna you are considering has an advertised gain of 3 *dBd*, your signal should be twice as strong as it would be if you were using a quarter-wave vertical. No antenna design can increase your power output; they can simply concentrate the signal in the horizontal plane, radiating less power skyward to wastefully warm the ionosphere. The law of reciprocity indicates you'll realize the same improvement on received signals as well. Not only will an antenna with higher gain increase the range of your signal, it will help to reduce the *picket fence* effect—a rapid chopping or fluctuation of signal strength often experienced by mobile stations.

As you compare advertised gain figures for various antennas, be sure you know the standard of comparison (dBd or dBi) being used. This is important when you remember a dipole has a theoretical gain of 2.14 dB over isotropic. Antennas referenced to an isotropic radiator have a gain figure of about 2 dB more than antennas referenced to a dipole. Although reputable antenna manufacturers strive to provide accurate gain measurements for their products, beware of imaginative claims! If the 19 inch long 2-meter antenna that has caught your eye is rated for 10 dB of gain, don't be surprised if it's 10 dBwdr—gain over a wet dish rag! Because of the difficulty in verifying advertised antenna gain, the ARRL does not print these gain figures.

Radiation Hazards and Antenna Placement

Most VHF/UHF antennas can be placed in several locations on your vehicle. Placing it in the center of the roof usually gives you optimum performance and the best radiation pattern. Aesthetic considerations, as well as the style of mount you are using, however, might prompt you to

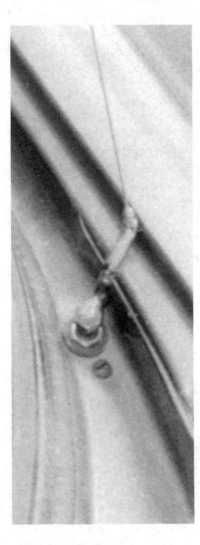

place the antenna elsewhere. If so, make sure you aren't exposing yourself or your passengers to dangerous levels of RF radiation. A cowl or trunk lip mounted antenna might be as close as 1 to 3 feet from the heads of your rear seat passengers. Mobile installations with a high power amplifier pose an even greater risk—the thought of 150 watts of 146 or 440-MHz energy warm-

Fig 5-1—David Brown, K8AX, built this almost invisible 2-meter antenna. A banana jack is mounted to the car body, just under the trunk lid. The antenna is made from piano wire and a banana plug. Worried about losing the antenna? Just open the trunk, unplug the antenna, close the trunk, and walk away. See the text for a discussion of antenna location and radiation safety. *(Photo courtesy of K8AX)*

ing your cranium isn't very comforting is it? The Safety Chapter section of *The ARRL Handbook* is mandatory reading for radiation exposure guidelines. Read it **before** you install a new antenna.

The Ubiquitous Mag-Mount

Magnetic mount or *mag-mount* antennas are an extremely popular choice for many hams. With a powerful magnet contained in the base, these antennas maintain a firm grip on the body of your car, eliminating the need to drill holes. In addition, their ability to be mounted and removed quickly makes them attractive to hams who frequently switch vehicles or who don't want to advertise the presence of their radio when they leave the car. Some new cars, however, cannot use mag-mounts because the car top is plastic or fiberglass.

If you choose a mag-mount antenna, a few simple precautions will minimize the possibility of damage to your car's finish. Before mounting the antenna, make sure the vehicle surface is clean and protected with a coat of good wax. If the antenna doesn't come with one, provide a protective pad between the antenna and the car—hams have reported success with various materials, including plastic sandwich bags, wax paper, even balloons. But be careful! Some materials can damage paint if left in contact for a long period of time.

If the material is too thick, it can seriously reduce the holding power of the magnet. Before mounting, check the base of the antenna to make sure it is clean and there are no burrs or protrusions—especially important if the antenna is pre-owned.

When placing the antenna on the vehicle, tilt the base and set one edge down first, then gently allow the remainder of the magnet to make contact. Once the antenna is in place, *never* slide it across the vehicle surface. When you remove the antenna, tilt it over onto the edge of the base and lift it free. Whenever possible, route the feed line lengthwise along the roof of the vehicle to the point of entry to minimize wind-induced movement that can scratch paint.

If your mag-mount installation is long term, it's a good idea to remove the antenna periodically and clean accumulated dirt and moisture from under the base. You may also want to reposition the antenna occasionally to lessen the possibility of paint discoloration. What appears to be discoloration may actually be where your antenna has shaded the vehicle surface from the sun's powerful paint-fading rays.

Glass Mount Antennas

If you consider a mag-mount antenna to be lacking in visual appeal, yet your conscience won't allow you to drill holes for a permanent antenna, a glass mount antenna may be the solution. Common among cellular phone users, they can work well on the amateur bands, too. If you own a sports car, the glass mount antenna can provide a reasonable compromise between aesthetics and antenna performance.

Although the concept of transmitting power to your antenna without a hard-wired connection might seem mysterious, the operation of a glass mount antenna is really quite simple. Two plates, one at the base of the antenna and the other on the end of the feed line, form a capacitor to couple the RF between the antenna and the feed line. Because this capacitor adds reactance that must be tuned out, a longer radiating element, or one with inductive loading—and sometimes a matching network at the end of the feed line—are employed to achieve a low SWR.

Build or Buy?

Many manufacturers make glass mount antennas, but given the simplicity of this antenna, why not build your own? If this idea appeals to you, you'll want to refer to two *QST* articles: "A Glass-Mounted 2-Meter Mobile Antenna," by Bill English, N6TIW, in the April 1991 issue, p 31, and "An Easy, On-Glass Antenna with Multiband Capability," by Robin Rumbolt, WA4TEM, in the March 1993 issue, p 35. Both articles have construction details and other useful information about glass mount antennas.

You will generally get best performance by mounting the antenna at the top center of the window, usually on the rear of the vehicle. Yes, you can mount your antenna at the top of the windshield, but be sure to check for windshield wiper clearance. One ham I know overlooked this vital step while constructing a glass mount antenna—his handiwork became a victim of windshield wiper assault the first time it rained. Besides allowing the greatest possible height above ground, placing the antenna at the top of the window provides the ground plane specified by some antenna manufacturers as necessary to achieve advertised gain. Remember, the nonmetallic roof found on some cars and motor homes will not act as a ground plane.

When mounting the antenna, avoid placing it over or in close proximity to window defogger/defroster grids or elements—these will deteriorate the antenna's performance and make it difficult or impossible to tune. You cannot use this type of antenna with the tinted windows (or some other coated window types) made with a metal layer.

If your antenna requires grounding of the coax shield at the antenna end, make the connection directly to the vehicle body, as near as possible to the antenna mount. Use a volt-ohmmeter to ensure the metal where you make your connection is indeed grounded to the vehicle body. Although glass mount antennas are removable, be sure you are satisfied with the location you have chosen before proceeding with the installation. If you find it necessary to remove the antenna, follow the manufacturer's instructions. Never use a metallic tool to pry the antenna base from the window.

Permanent Mount Antennas

Glass-mount antennas have a trendy appearance that is truly chic, but their method of attachment to the vehicle limits them to a relatively short radiating element—typically a quarter wavelength. If you need better performance from your mobile antenna than a glass mount can offer, you will probably want to consider a permanent mount antenna. In addition to providing superior performance, when it comes to appearance, permanent mount antennas are hard to beat.

Although the term "permanent mount" is a bit of a misnomer (obviously, you *can* remove a permanent mount antenna), it's generally used to describe any antenna requiring a mounting hole. The decision to take a drill to the body of your auto can be a difficult one. Consider it carefully before drilling. Aside from affecting resale value, drilling a hole in your auto can set the stage for future problems with rust—and provide water with a point of entry into the vehicle. Proper installation and maintenance are essential.

As mentioned previously, the antenna should be mounted in the center of the roof, which provides the largest, most uniform ground plane. On some vehicles, this location will be over the dome light. When you remove this light, you will have access to the lower side of the roof. Truck owners often locate the antenna at rear center of the roof to take advantage of dome light placement in those vehicles. Many mini-vans will give you convenient access by removing the overhead console. On vehicles with a reinforcement channel located in the center of the roof, you will probably want to shift the location of the antenna slightly to avoid having to drill the channel.

Take careful measurements and double check them before you drill. If your vehicle has a retractable sunroof, determine its area of travel before choosing a location for your antenna. If the antenna won't be installed over the dome light, it is a good idea to detach and lower (or remove) the headliner before you begin, providing you more room to work *and* possibly preventing damage to the headliner. Drill bits and hole saws have a voracious appetite for headliner material—you don't want them to indulge in a feeding frenzy on your vehicle.

Exert light pressure when drilling the hole, especially if you are using a hole saw. The thin metal roof of most modern automobiles is easily deformed, especially if the drill breaks through unexpectedly while you are balancing your full body weight on the handle. If you have access to a *chassis punch*, use one; it makes a neat and precisely dimensioned hole. De-burr the finished hole, and lightly feather the paint along the edge of the opening with fine sandpaper to prevent chipping and subsequent rust. As you complete the installation, be sure to install the weather seal or O-ring provided with the antenna.

If you aren't comfortable with the idea of doing your own through-hole antenna installation, you may want to enlist the services of your local two-way radio shop. Most of them will be glad to mount your antenna for a nominal fee, relieving you of the work and worry.

How Many Ways Can You Skin a Cat?

As soon as you install a permanent mount antenna, your car's resale value will plummet—or will it? Depending on the vehicle, the antenna and your savvy, perhaps not. If you are concerned about the effect of a permanent mount antenna on the value of your automobile, consider the following tips:

Plug the Hole

If your antenna uses a standard size mounting hole, your local two-way radio shop can provide a snap-in rubber plug made just for filling antenna holes. If your vehicle is tall (a van or truck, for example), it's possible the plugged hole won't be visible to anyone who's less than 7 feet tall. This isn't a suggestion to deceive—it's just that some prospective owners of your vehicle won't object to the plugged hole if it isn't readily visible.

Patch the Hole

Before you drill the hole for an antenna, check with your dealer or a body shop to see what the charge will be to fill the hole and repaint the surrounding area at trading time. Often the charge is less than the deduction in vehicle value that the dealer or prospective buyer will assess.

Remove the Hole?

My friend Mark, WA4JSN, was faced with this soul-searching decision: Should he put a hole in the roof of his new van? With drill in hand, he was poised ready to do the "deed" when he spied a more palatable solution. On his particular van, the center brake light is mounted on the roof in a removable

Fig 5-2—While his antenna is not placed optimally in the center of the roof, Mark Goff, WA4JSN, did not have to drill a hole in his car's metal roof. He used the roof-mounted stoplight (see text). At trade-in time he will simply replace the light housing. (*Photo courtesy of WA4JSN*)

metal housing. After removing the housing, he found there was both ample room to mount his antenna and the hole for the existing light wires was large enough to allow passage of the coax.

Before the antenna was mounted, a trip to the local salvage yard netted him a spare housing, to be installed when he trades vehicles. Even though the antenna might perform better if it were in the center of the roof, the compromise location (Fig 5-2) has proved to be more than adequate.

Give 'em the. . . Antenna?

Having a cellular phone in your car has become quite the status symbol. So much so that some companies offer dummy cell phone antennas to those who want the image without the expense. What does this have to do with drilling holes for Amateur Radio antennas? Plenty. Check your antenna catalogs and you'll find many manufacturers make cellular antennas using the same type mount as their Amateur Radio products. When it comes time to trade or sell, replace your ham antenna with a cell phone antenna and be sure to advertise your vehicle as "cellular ready."

Trunk Lip Mounts

You say you want the good looks and the performance of a permanent mount antenna, but without drilling holes? The trunk lip mount may be your answer. Available in a variety of styles and configurations, the trunk lip mount offers a sturdy and attractive support for the largest VHF/UHF antennas—many models also allow the antenna to be folded over for clearance when necessary.

Placement of a trunk lip mounted antenna is often governed largely by aesthetics, but keep in mind mounting it in the center of the forward edge of the trunk lid will skew the radiation pattern and possibly make tuning difficult—due to the close proximity of the roof. Cars with a hatch fare better in this regard since the antenna can be mounted near the top of the hatch, clear of surrounding metal. When contemplating the use of a trunk lip mount, take into consideration the amount of radiation rear seat passengers may be exposed to—as previously mentioned.

Installation of a trunk lip mount is straightforward—but follow the manufacturer's instructions carefully. Remember, the trunk lip antenna depends on a good ground connection to the body of the vehicle for proper operation. This means you will have to remove paint where the mount's setscrews contact the trunk lip. Mark contact points carefully, and remove the minimum area of paint necessary to allow setscrew-to-trunk lid contact. This is a good time to use your VOM to determine if the trunk lid is electrically connected to the rest of the vehicle.

Before installing the mount, coat the bare spots of metal with a good conductive anti-seize compound, and be sure to periodically check the mounting point for any signs of rust. With the antenna mounted, carefully open the trunk or hatch while observing for clearance problems between the antenna and the vehicle body. On some vehicles it may be necessary to use an antenna or mount with fold-over capability.

Trunk Gutter Mounts

No relation to the trunk lip mount, the trunk gutter mount requires two or three small holes to be drilled to facilitate installation. Fortunately, because the holes are made in the gutter under the trunk lid, they aren't visible when the trunk is closed. When the time comes to remove the antenna, the small holes are easily filled, requiring no expensive body and paint refinishing.

Contrary to their name, they aren't limited to trunk

gutter mounting, but can be used just about anywhere there is a gap in body panels. Due to the simplicity of the trunk gutter mount, if you are handy with metal working tools, you might want to fabricate your own. If the mount requires you to drill holes, the finished installation should be checked frequently for the appearance of rust or water leakage into the vehicle.

Fig 5-3—A trunk gutter mount is easy to install. As with any rear deck mounted antenna, the radiation pattern will not be symmetrical. It will favor some directions and not work as well in others.

HF Antennas

Everyone knows the most important rule of HF antennas: Make them big, and put them high in the sky. This is quite a challenge for the prospective mobile HF operator.

After all, with the possible exception of 10 meters, your HF mobile antenna will always be shorter than a quarter wavelength—and it certainly isn't going to be very far off the ground. Since mobile HF antennas are such a compromise, their performance will always rank somewhere between poor and terrible. Right? Well... don't tell that to my friend Dale, NU4O. He has worked more than 100 countries while mobile, running only 50 watts.

Choosing the *Right* Antenna

Your choice of antenna will be influenced by many

considerations, such as appearance, cost, performance and the ability to QSY—within a band and from band-to-band. Several manufacturers offer mobile HF antennas—or perhaps you'll want to build your own.

Monoband Antennas

If your plans for mobile operation don't include a lot of band-hopping, you may want to consider using a monoband antenna. Generally, they are cheaper, simpler and somewhat more pleasing to the eye than their multiband cousins. They come in two basic configurations.

In a design made popular by the Hustler antenna, a fixed-length mast is topped by a removable loading coil-whip section, known as a *resonator*, which allows the antenna to resonate in a particular band. In addition to the standard resonator, the Hustler antennas are available with "Super" resonators, which provide greater power handling capability and more bandwidth. Because you only purchase one resonator for each band you plan to operate, you won't be spending money for extra band capability you don't require. If you frequent areas with low clearance, you'll appreciate the Hustler's fold-over mast.

Another approach is to make the entire antenna function on one band only. Some brands, such as the Ham Stick and the Jetstream, utilize a fiberglass mast helically wound with a copper radiator and topped off with an adjustable stainless steel whip. With performance comparable to the Hustler-type antenna, many hams choose this style of antenna for its sleeker appearance.

If aesthetics are a top priority, check out the line of diminutive monoband antennas made by Comet. Offered in versions for 10, 20 and 40 meters, they consist of a short, stainless steel whip with a center loading coil. How short? The 40-meter version is only 40 inches in length.

No matter what brand you choose, you'll need separate antennas for each band you plan to operate.

Multiband Antennas

Since you'll probably be using a multiband rig, why not complement it with a multiband antenna?

As the name implies, multiband antennas can be operated on several of the HF bands, with most models covering 10-80 meters. In addition to appearance and size, the method of bandswitching will be a factor in your choice of antenna.

Manual Bandswitching

If you don't mind having to get out of your car to change bands, you have a choice of several multiband antennas. One very popular design is the Bug Catcher antenna, marketed under several trade names, including the Texas Bug Catcher and the Carolina Bug Catcher. Believed to have been developed and named by the Texas Army Corps of Engineers in the 1920s, the Bug Catchers use a tall mast with a large, air wound center loading coil, topped off with a capacitance hat and stainless steel whip. Band changes are made by adjusting the location of a tap on the loading coil. While the Bug Catchers are considered to be top performers, their imposing presence, especially when used on smaller vehicles, might be aesthetically stressful. These antennas are heavy and their size subjects them to quite a bit of wind loading, so they'll need a high strength mount and possibly guy rope or twine to prevent excess movement.

Since the construction of the Bug Catcher is readily apparent to anyone who looks at one, many hams choose to build their own version, incorporating changes or modifications as they see fit.

Generally acknowledged as one of the most rugged multiband antennas on the market, the Terlin Outbacker antenna is appearing on vehicles all over the world. Offered in several versions in lengths up to 7.5 feet, the Outbackers cover 10-75 meters with power ratings up to 300 watts. The antenna's two-piece construction consists of a fiberglass mast—wound with a helical radiator covered with an epoxy resin coating—and an adjustable stainless steel whip. Band changes are accomplished by connecting a jumper (the "Wander Lead") to the appropriate tap on the mast. Hams report good performance from the Outbacker, in addition to its reputation for seldom losing a fight.

Automatic Bandswitching

It must be my "Type A" personality. When I decide to change bands while mobile, I want to do it instantly—no delays, no stopping to get out of the car and shuffle antenna parts around. For me, a multiband antenna with automatic bandswitching is the *only* way to go.

Fig 5-4—John Gerlach, K6BRD/7, built this HF antenna array. It consists of a standard 52-inch Hustler fold-over mast and a home-brewed spider. John built coils for 17, 15, 12 and 10 meters, each with a capacitance hat. The tennis balls keep the mast from banging into the car body, and the bungie cord restrains the mast from tilting back at highway speeds. *(Photo courtesy of K6BRD/7)*

Although hams have tried many forms of automatic antenna bandswitching, the most popular method uses *multiple resonators* in what is often referred to as a "Spider" design. The spider uses a resonator for each band, spaced radially around an adapter attached to the top of the mast. Because the resonators for the bands you *aren't* operating don't present a match for your rig, they are electrically divorced from the system and don't accept any power from the transmitter.

Several companies make spider-style antennas. Multi-Band Antennas makes the original Spider, or if you own a Hustler antenna, they can supply an adapter to mount several resonators at once on the Hustler mast—a la the spider.

Another very popular antenna you may have seen or heard about is the *screwdriver* antenna. Although in its basic form it doesn't automatically change bands, it can be adjusted from the driver's seat to operate *anywhere* within its designed frequency range, usually the 10 to 80-meter bands. Controllers to automatically resonate the antenna are available.

The brainchild of Don Johnson, W6AAQ, commercial versions are advertised in *QST* by MB Products, High Sierra Antennas and the T. J. Antenna Company.

Hams who have tried the antenna give it high marks for performance and convenient operation. Although the construction of the antenna requires moderate fabrication skills as well as access to some machine tools, it is a popular homebrewed HF mobile antenna design.

Spanning the Band

At 14 MHz and above, most mobile antennas have sufficient bandwidth to allow reasonable frequency excursions without the necessity of retuning. At lower frequencies, antenna bandwidth diminishes, and even moderate changes in frequency can cause SWR to rise to a level your rig won't accept. If you desire the ability to cover much or all the bands you plan to operate, what are your options?

For an antenna tuned to resonance with an adjustable whip, you can use a permanent marker to identify the points on the whip corresponding to resonance, on the frequencies you plan to operate. This will allow you to quickly change the resonant frequency of the antenna without the need for an SWR meter. Likewise, antennas using tapped coils for tuning can be marked for the appropriate tap locations.

But what if you'd rather not have to stop and make adjustments just to move from one portion of the band to another? In this case, you can use an antenna tuner or Transmatch to achieve a low SWR when you are operating your antenna at a nonresonant frequency. If you haven't yet selected a mobile rig, you may want to consider one of the many models containing a built in tuner. Not only will this save you valuable space inside your vehicle, internal tuners adjust themselves automatically—all you do is key the rig and the tuner quickly achieves the best possible match. If your rig doesn't have an internal tuner, small outboard tuners, suitable for mobile use, are available from several manufacturers.

Make all adjustments to an outboard tuner when the automobile is stationary—watching the SWR meter while you tweak the tuner and drive is dangerous. Don't try to use a tuner to make your 10-meter antenna work on 75 meters. Not only will increased loss make the antenna very inefficient, resonators can get very hot when force fed out-of-band RF, possibly sustaining permanent damage.

Mounts

Perhaps one of the biggest challenges faced by the would-be mobile HF operator is how to attach the antenna. Back in the days when cars had real bumpers, it was a simple matter to bolt on a spring base, or clamp on a bumper strap mount. No more. So what's a ham to do? Let's have a look.

If you are totally dedicated to HF mobile, the optimum installation can be had by drilling a hole and mounting the antenna in the middle of your car's roof. And if you are using one of the really large antennas, you might be able to get the job of trimming low tree branches along the streets of the your city. Seriously, in addition to the possible effects on the property of others (one ham reported his roof-mounted Outbacker antenna shattered fluorescent light bulbs under the canopy of his local gas station), consider the antenna's effect on the area where it is mounted. You may need to provide some type of reinforcement to prevent fatigue and fracture of the metal surrounding the mounting hole.

If you're squeamish about drilling holes, there are other ways to get your antenna on the roof—and keep it there. If your vehicle has a luggage rack, there are several companies that make mounts that clamp to the rack, or you can build your own luggage rack mount from some flat stock.

Be sure the rack is grounded, or add a good, low impedance ground connection from the antenna ground connection to the body of the vehicle. What's low impedance? A friend of mine reported his 20-meter mobile setup had high SWR and the mike was hot with RF. It turns out he had fabricated a drop-in stake hole mount to use on the bed of his pickup. Since he didn't want to drill a hole to connect the ground strap, he had routed it down to the chassis of the truck. The inductive reactance of his so-called ground strap was more than 100 ohms at 14 MHz.

Although mag-mount antennas were once considered to be practical only for VHF/UHF antennas, many hams now use them for HF antennas as well. Utilizing three or four powerful magnets attached to an "H" or "T" shaped metal framework, mag-mounts provide a no-holes required,

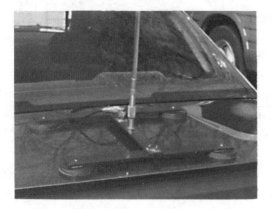

Fig 5-5—With a little care and thought, HF antennas can be either mag mounted or roof-rack mounted. John Bartucci, K8JB, built this four-magnet mag mount for the rear deck of his car. *(Photo courtesy of K8JB)*

stable, wide-stance support for even the largest antennas.

Magnetic HF antenna mounts are available from the Lakeview Company and from Metal and Cable Corporation. If you prefer to build your own, refer to an article by Ed Karsin, W3BMW, titled, "The Impossible Dream Whip HF Mag Mount," in the December 1992 *QST*. Remember, while the large area of the mount provides a capacitive electrical ground for the antenna, many installations work better with a direct ground connection to the vehicle.

Refer to the previous discussion of mag-mount VHF/UHF antennas for installation and maintenance information. Remember to test the mount for stability before you head out onto the freeway—if dislodged, one of these mounts with an antenna attached can do a great deal of damage to your vehicle.

Mounting the antenna on the rear of the vehicle is a more attractive option for many hams, but rubber-covered bumpers offer no support for this approach. Well actually, there is an honest-to-goodness metal bumper lurking beneath the rubber cover. But since the bumper cover is usually not repairable, and replacing it is quite expensive, punching it full of holes just to mount an antenna isn't very practical. Fortunately, you may not have to. If you examine the lower edge of the bumper, you'll probably find one or more holes—sometimes where push-in plastic retainers are used to hold the rubber cover against the bumper. Of course, bumpers vary, but you can almost always fabricate a bracket—flat aluminum stock works well—that will extend beyond the edge of the bumper to provide a mount for your antenna. Even if you have to drill holes, they won't be visible to anyone but ground-based insects and animals. Since the bumper is mounted on shock-absorbing brackets, it's a good idea to install a short ground strap from the chassis to the bumper to ensure a good electrical connection.

If you are unable to attach a mount directly to the bumper, it may be possible to use a longer bracket, attached to the frame and bent where necessary to clear the bumper. Either way, be sure your installation doesn't interfere with the operation of the bumper shock absorbers.

You also can mount an antenna to some part of a trailer hitch assembly. Although some hams mount their antenna in the existing hole provided for the hitch ball, this selection obviously limits your HF mobile activities to times when you aren't towing a trailer. If you mount the antenna to the slide-out section of a *receiver* hitch, you'll need to provide a ground strap from

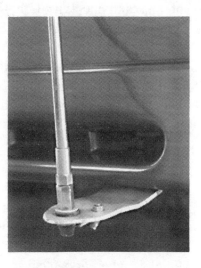

Fig 5-6—Pick up your antenna mounting point under the bumper. This plastic covered bumper is bypassed by a piece of sheet aluminum. Notice the ground lug just to the right of the whip mount.

the hitch to the vehicle chassis to guarantee proper antenna performance.

A trunk lip mount may be adequate for one of the smaller HF antennas,. Consult with the antenna manufacturer for recommendations, and refer to the VHF/UHF antenna section for installation guidelines.

Antennas After Installation

The Care. . .

Life on the road is no joy ride for your mobile antenna. Wind loading, continual flexing, vibration and occasional blows from low obstructions can test the mettle of the best of antennas. Be sure to provide a secure mount for your antenna, and check it often—especially after antenna/low-flying object altercations. Some antennas use a whip-retention setscrew or collet. You should periodically check these screws for tightness. Having your antenna become a hood ornament for the car behind you on the freeway can have costly and potentially disastrous consequences.

. . . Feeding

Feed your antenna with high quality coax—preferably one with at least 95% shield coverage. Even though most mobile feed line runs are relatively short, the higher quality coax will ensure that more of your rig's RF arrives at the antenna, and less of it ricochets around the inside of the vehicle—where it can cause automotive electronic systems to behave rather mysteriously.

As mentioned in Chapter 4, take care to route the feed line away from vehicle wiring or electronics. If you must traverse a wiring harness with your feed line, make the crossing at right angles only. Feed line for rear mounted antennas is usually best routed through a hole in the rear of the trunk or floor pan.

Check to see if you can use the existing plugged body holes—most vehicles have them. Feed line routed under the vehicle should be pulled snug with no excess slack left dangling. Secure it with cable ties or clips and use rubber hose as a grommet where it passes through holes in the frame. Be sure to avoid exhaust and suspension components. Avoid making sharp turns or bends that deform the cable—these may cause shorter life and possibly increased SWR by altering the characteristic impedance. With few exceptions (such as magmount and some glass-mount antennas) the coax shield should be securely grounded to the vehicle body or chassis as close as possible to the antenna. Remember to carefully seal the exposed end of the cable to prevent water migration between the jacket and center dielectric.

If your installation necessitates bringing the feed line into the vehicle through a door or the trunk, choose your point of entry very carefully. Although most doors and trunk lids can be safely closed on small (RG-58 style) coax without causing cable damage, the adjustment of latches and the thickness of weather stripping will be the determining factors. Besides causing high SWR, flattened cable can become shorted. Rule of thumb: With the door or trunk closed, you should be able to pull the cable through the gap with only a moderate tug.

Routing cable through the top of a rear hatch requires extra caution since the lip of the hatch usually recedes below the roofline as the hatch is opened, making a very effective feed line shear. (I once severed my mag-mount antenna's feed line when it shifted a couple of inches as I was raising the hatch on my car.)

. . . and Tuning

The finished installation should be tuned for the lowest possible SWR—to maximize performance, and to minimize

Fig 5-7—Here is a more classic HF antenna mount. An 8-foot steel whip with a base loading coil is used by Steve Cerwin, WA5FRF. Steve does have an advantage with his Ford Bronco—the fiberglass top does not affect the performance as a metal top would. *(Photo courtesy of WA5FRF)*

interference to automotive electronics. Although it may not be possible to obtain a perfect match, try to achieve an SWR of 2:1 or better, especially on VHF/UHF antennas. Adjust your antenna system with the vehicle in a clear location, away from power lines, trees and buildings, which can affect tuning. Make sure the trunk lid/hatch and doors are closed while you take SWR measurements. Use an accurate SWR meter, designed for the band to be checked. Using an SWR meter outside its designed frequency range will give erroneous power and SWR readings.

Before you put the meter away and declare the installation finished, with an assistant driving, take your measurements again with the vehicle in motion. Although moderate fluctuations in SWR are normal as the antenna flexes and passes nearby objects, severe changes may reveal problems with the antenna or the need to reengineer your installation.

CHAPTER 6

Operating Mobile: On the Road, On the Air

ou're going to love going mobile. No longer will you have to cut short a good conversation because you have to be on your way somewhere. You can take the conversation with you. And should an emergency arise while you are on the road, your "hobby" might even save a life.

Speaking of emergencies, don't let your radio operation create one. If traffic or road conditions demand your undivided attention, stay off the radio. If you are involved in a conversation and you realize your ability to drive is compromised, either pull off the road or sign with the station you are working. No QSO is worth risking your safety or that of others.

Hitting the Road on VHF/UHF

Cars and VHF/UHF operation are made for each other. If you're sprinting around town or crisscrossing the country, the noise-free convenience of FM VHF/UHF radio is hard to beat. Especially with all those repeaters out there.

Getting to Know your Rig

A sophisticated mobile rig can be a real pleasure to operate—once you learn how. But there's nothing that says a mobile rig can only be used mobile. Try operating your new rig from home for a few days before installing it in your car. This will give you a chance to become more familiar with the rig's controls, which will make your mobile operation more enjoyable—and safer.

Getting Maximum Mileage from Repeaters

When you are out on the road, perhaps one of the most discouraging things you can hear is the dreaded sound of silence right after you've keyed a repeater and announced, "KB4AEY mobile, listening." Communicating—that's what ham radio is all about. So why isn't someone out there just jumping at the opportunity to communicate with you?

Well first of all, you shouldn't take it personally. It has nothing to do with halitosis. Although several hams may have heard your transmission, they are waiting to see if someone else is going to greet you. After an extended period of silence, they may even conclude you aren't really serious about talking to someone. (Or possibly they've been burned by a rude operator. See the sidebar, "Where's This Repeater Located?") How do you increase the odds of someone answering you? Ask any tournament-winning fisherman the secret to his success, and he will tell you that to catch fish, you've got to use the right lure.

If the repeater you are monitoring has a conversation in progress, you might want to use a technique known as "tail-ending." That simply means calling one of the participants when the conversation has ended. Make sure you don't call an operator who has indirectly indicated he won't be listening for further calls, for example, "I've got to jump

Fig 6-1—On the road, or really on the trail, the Sparks family—D.D., KN4EC and Reba, KC4IPO, keep in touch with the outside world. (*Photo courtesy of KN4EC*)

out and fill up with gas. 73 from KA6WAR." (Calling the recorded voice ID of a repeater is also not considered good practice.) Of course you can join an ongoing QSO by interjecting your call sign between exchanges. Transmitting "break" is usually reserved for emergencies on most repeaters. Be considerate of the other operators, and make sure you can add something meaningful to the discussion.

But let's suppose you've been monitoring a repeater for some time and have heard nothing but the occasional "kerchunker" and repeater identification. (Not at all unusual for repeaters located in rural areas.) What now? Since the old standard, "WF4N, mobile monitoring," will probably kindle about as much action as holding a lighted match to a soggy newspaper, why not spice things up a bit?

"This is WF4N mobile from Kentucky, entering

"Where's This Repeater Located?"

Regardless of whether your journeys take you across the state or across the continent, you'll find *The ARRL Repeater Directory* an indispensable addition to your mobile station. At your fingertips you will have frequencies, locations, call signs, and other pertinent information (CTCSS tone frequencies, for example) for every repeater in the US. When planning a trip of any distance, I like to mark the appropriate points on my atlas with the frequencies of some of the repeaters listed in the directory. I make my marks with a highlighting pen to make it easier for Judy, my wife and chief navigator, to read road information.

Of course you could save a few bucks by not having a repeater directory and just asking someone the location of each repeater you hear. No big deal, right? Well, consider the following scenario.

You're at the workbench, using a toothpick to gingerly coax a surface-mount component into place, your other hand holding a soldering iron poised ready to anchor the little critter to the board. Just as the marriage of component and solder is about to commence, a voice bursts forth from the 2-meter rig on the other side of your shop. "This is KB4AIZ, is there anyone listening?" You pause for a moment. "Surely someone will answer him," you mutter under your breath as you refocus your attention on the task at hand. A few seconds later, "This is KB4AIZ, is there anyone listening on this repeater?"

Sensing a distinct note of urgency in the caller's voice, you put down your tools and go to your rig. "This is

Flagstaff on I-40 and looking for someone to chat with. Is anyone around?" Similar to extending a hand of friendship when meeting someone in person, a brief introduction of yourself to prospective contacts is an important first step in breaking the ice. Making it known that you are from outside

WF4N. Is there something I can do for you?" you ask.

"WF4N, this is KB4AIZ. Could you tell me where this repeater is located?"

Okay, so your first inclination is to tell him that the repeater is located in the little building at the base of the tower. But your mother taught you better manners. . . So you answer, "Sure, it's located in Stroud Station, Kentucky."

"Oh, okay. I've worked this machine several times as I traveled through the area, but the courtesy tone is different tonight and I didn't recognize it. Thanks for the info. KB4AIZ, clear."

"Sure, no problem," you say, trying to remain diplomatic, "glad to be of assistance. WF4N clear."

With your blood pressure rising like an Apollo rocket booster, you return to the workbench to find the component you were courting has eloped with a dust bunny, leaving no forwarding address. But at least that fellow knows where the repeater is located.

With voice IDers becoming commonplace (and of course, with a repeater directory in hand), the traveling ham should never have a problem determining what repeater he is hearing. The rare commodity called patience helps too. Most repeaters ID every six or seven minutes when in use. In rare instances, where you cannot identify a particular repeater, by all means ask someone. The locals are always glad to make you aware of the features and coverage of their machines. But be courteous and invest a little time in some good conversation in return for the information you receive.

the area tends to pique the interest of the "traveler spirit" that's intrinsic in most hams.

If you are planning to spend some time in the area, you might want to expand your invitation just a bit.

"This is WF4N mobile from Kentucky. We're in San

Diego for a couple of days and I'm wondering if someone can recommend a good restaurant."

Not only have you introduced yourself, you've given your fellow hams an opportunity to do what hams like to do most: Help others.

Preserving Your Welcome

Good repeater etiquette is important. Remember—when you use someone's repeater, you are a guest. So be polite.

One of the easiest mistakes for the mobile ham to make is that of monopolizing a repeater. When you are engrossed in a good philosophical QSO with a ham you've just met, the miles can really fly by. And so can the minutes, or if the repeater has exceptional range, maybe even hours. Just like the party line telephones of yesteryear, when you are using the repeater, everyone else waits.

Once you've established contact on a repeater, if possible, you and the other station should move to a simplex frequency. (The REVERSE button on newer rigs is specifically included to allow you to monitor the repeater input frequency while the other station transmits. If you can receive him on the input frequency, you should be operating simplex. Simple, no?)

But what if you can't work simplex? No problem. Repeater owners welcome the use of their machines by mobile operators. After all, this is the primary purpose of a repeater— to extend the range of mobile and portable stations. But be sensitive to the need of other hams to use the machine, especially during "busy" times of the day or evening, for example, rush hour and lunchtime. And don't forget to identify regularly.

If you are traveling alone, ham radio is a great way to combat the boredom of a long trip. But don't let your conversation grow bore-some to everyone listening. How many times have you heard an extended monologue such as this one?

"Okay, yeah the traffic is pretty heavy out here today, uh, there goes one of those uh, you know, big motor homes and he uh, he sure must have a lot of money in that rig, and uh, yikes, some @%! guy just cut me off and uh. . ."

Avoid the tendency to ramble. If you can't think of anything significant to say, leave your mike unkeyed.

Keep the volume of your car stereo at a reasonable level while you are transmitting. If the folks on the frequency want to party, let them provide their own music.

With Some Help From Friends. . .

Although repeaters have sprouted up like magic mushrooms all over the country, be assured it takes a large investment of labor and money to put a machine on the air and keep it there. If you find yourself frequently using a particular repeater, why not help defray operating expenses and show your appreciation for the availability of the machine—by joining the sponsoring club or repeater group? Your contribution will be greatly appreciated.

Crossband Repeat and Remote Control

As mentioned in Chapter 3, enlisting the crossband repeat and remote control capabilities of your mobile rig can greatly add to your operating pleasure. By following some simple guidelines, you can keep your operation fun and legal.

The appropriate selection of operating frequencies for crossband repeat is important. First of all, you want to make sure you don't choose harmonically related frequencies. As an example, you program in 147.50 MHz for your mobile rig's transmit frequency and 442.50 MHz as your HT's transmit frequency. As soon as you key your HT on 442.50 MHz, your mobile rig receives it on this UHF frequency and starts to retransmit your signal on 147.50 MHz. Problem is, when you unkey your HT, the mobile rig's VHF transmitter will continue transmitting because the third harmonic of its 147.50 MHz output is keeping the squelch open on its own UHF receiver (listening on 3×147.50 or 442.5 MHz). As if that isn't enough, it will be transmitting a tremendous howl due to audio feedback. Depending on the rig, you may find the only way to rescue it from its locked-up condition is to turn it off, connect a dummy antenna, turn the rig back on and take it out of crossband repeat mode.

It's also important to avoid the input, output, link and control frequencies of fixed-site repeaters in your area. A repeater directory and some listening will identify the input and output frequencies, but the link and control frequencies are a little trickier since repeater owners wisely choose not to publicize them.

Considering that all repeaters (even mobile ones) should be coordinated anyway, your best bet is to contact your local frequency coordinator for assistance in choosing the proper frequencies for crossband repeater operation in your area.

Crossband repeat is useful and fun, but it isn't without its drawbacks. Fortunately, most are avoidable.

If you use your crossband repeater to relay your HT's signal to another repeater (one on 2 meters, for example), consider the effect of that repeater's *hang time* on your operation (hang time is the period that the repeater stays keyed after it has stopped receiving a signal—typically 2 to 6 seconds in duration). Because a crossband repeater will retransmit the first signal it receives for as long as it receives it—you might find it hard to get a word in edgewise on a busy repeater. If you are able to hear the repeater's output directly on your HT, you can configure your crossband repeater for fixed direction crossband repeat. This simply means your crossband rig will only repeat the signals received on one band (the band where you are transmitting

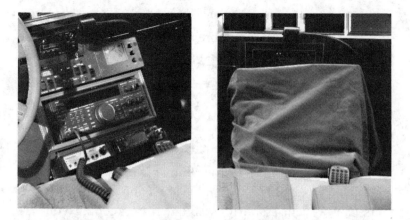

Fig 6-2—While operating, Mark Lesar, N8KOB's rigs are quite visible. For security, when parked, a piece of cloth covers the equipment. Normally the mike would be tucked under the cover. (*Photo courtesy of N8KOB*)

with your HT). This will make it possible for you to transmit without waiting for the 2-meter repeater to unkey.

When using your rig in crossband repeat mode, use the lowest power settings on both bands—that will provide reliable communications. This will not only help to prevent you from returning to a vehicle that won't start because of a dead battery, it also will reduce potentially damaging heat build up in your rig. (It also lessens interference on the bands.) Don't forget, anytime it's receiving, it's transmitting. If you have programmed a busy frequency, your rig may be transmitting non-stop for hours.

Of course if you have the ability to remote control your crossband repeater, you can reprogram a different, less active frequency. Since the FCC considers the remote control of one radio by another to be *Auxiliary Operation,* you must follow the applicable rules. Primarily, this means your control signals must be transmitted within the amateur bands on frequencies above 222.15 MHz, with the exception of the segments 431-433 MHz and 435-438 MHz. (Interestingly, since my homebrew remote car starter is not a radio, I can transmit control signals to it via the receiver in my car's 2-meter rig.)

As the control operator, you are responsible for what is retransmitted by your crossband repeater. To prevent unauthorized use, most rigs allow you to utilize CTCSS tones or a DTMF tone sequence to activate a crossband repeater. If your rig has a programmable time-out-timer, it's a good idea to enable it when you are using the rig to repeat. This can prevent a spurious signal from keeping your rig keyed until it reaches the melt-down point.

Identifying with your HT is adequate when your signal is being repeated, but it doesn't cover your mobile rig when it is repeating signals back to you. To be legal, you'll need to install an IDer in your crossband rig.

HF Mobile

I'll never forget my first experience with HF mobile. I was a young lad, and with the expectation of becoming a ham very soon, had just purchased my Novice station, a second hand Johnson Viking Adventurer transmitter and Hallicrafters SX-99 receiver. Naturally, I was spending a lot more time listening to the '99 than those dreadful 78 RPM Morse code practice records.

One evening, as I tuned across the 75-meter band, I happened upon an interesting conversation between a couple of guys who were using AM. (Although AM was quickly becoming a thing of the past by the mid-'60s, many hams still preferred the mode. My lack of radio experience coupled with the '99's instability—made all SSB operators sound like Donald Duck, so I also preferred AM.)

It turned out one of the hams was mobile, en route from his home in a small New England town to his winter home in

Florida. I was fascinated. This fellow was able to talk to other hams around the country as he traveled down the highway. For me, it was better than a Dick Tracy wristwatch-television. I listened for several hours as he spoke of the towns he was passing through, the sights he was seeing, the stops for gas, doughnuts and coffee. During those hours, which stretched far beyond my "official" bedtime—thank goodness for headphones—I was an invisible passenger with that traveling ham, enjoying the trip as though I were there.

Many years have passed since that enchanted evening, but the thrill of HF mobile is as strong as ever. Of course, now I'm the one who's hauling those unseen passengers along as I travel; and is it ever fun. You can get in on the fun too. HF and the road are calling.

Netting Contacts

Just getting on the air and calling or answering a CQ is a sure-fire way to make lots of HF contacts. But if you want to add a little variety to your operating, explore the various mobile nets in operation on the HF bands. The wide area nets, MIDCARS on 7.258 MHz , EASTCARS on 7.255 MHz and SOUTHCARS on 7.251 MHz perform a dual purpose. In addition to monitoring the designated frequency for calls from mobile operators in need of assistance, these nets serve as a meeting place of sorts for the traveling ham. Let's look at an example of how they work.

Suppose I'm traveling from New York to Texas and I want to periodically let Joel, N1BKE, in Connecticut, know how my trip is going. At the prearranged time, I call the MIDCARS net control station with a request to make a contact. Permission granted, I then give Joel a call. If he answers, we move to another frequency to continue our QSO.

But what if Joel is working late and isn't able to meet me on time? In that case, I can leave a message for him on

Mobile Amateur Radio Awards Club

Interested in doing a little contesting while mobile? The Mobile Amateur Radio Award Club sponsors several awards, the most popular being "Worked All Counties." Their net frequencies also act as a convenient place for mobiles to meet and get help if they have any problems. For more information, contact Fred Crawford, AB5SL, at PO Box 2561, Universal City TX 78148-2561.

the net, letting him know where I am and when I'll call him again. Later, when Joel checks in looking for me, the net will deliver my message to him.

Or it could be Joel's on time, but poor propagation won't allow us to hear each other. If this happens, it's almost guaranteed at least one of the stations on frequency will be able to hear the both of us and will perform a relay, enabling us to exchange necessary information, for a move to another band perhaps.

Of course you don't have to be keeping a schedule with someone to get a lot of operating enjoyment from nets. You'll receive a hearty welcome to the County Hunter's nets on 3.865, 14.056 and 14.336 MHz. You've heard of WAS (Worked All States)? Well, the county hunters cut the WAS pie into infinitely smaller pieces by attempting to work someone in every county in the US. Since many of the more than 3000 counties in the country don't have resident hams, the county hunters rely on mobile hams to make those counties available for contacts. If you are willing to detour slightly from the beaten path, the county hunters will reward you with a gigantic pileup.

Oh, and don't worry if you have no idea what county you are traveling through, because these hams take their county-hunting business seriously. If you can provide a road number and some landmark information, they will tell you where you are.

Keep in mind, the frequencies used by the County Hunter's nets are usually monitored by the Mobile Emergency net. If you suffer the misfortune of having car trouble, someone may be available to summon aid for you.

There are literally hundreds of nets on the HF bands, each catering to a particular interest. RV mobile operators will want to check out the Good Sam RV and Family Motor Coach Association nets. Or you might want to check into one of the myriad of nets having nothing at all to do with mobile operation. The choice is yours.

Times and frequencies for nets can be found in *The*

Fig 6-3—Harry Wheeldon, KM4C, doesn't let hay-cutting season interfere with his ham radio activities. He goes tractor mobile. Notice the through-the-glass antenna on the windshield. (*Photo courtesy of KM4C***)**

ARRL Net Directory. Take a copy with you when you travel; net operation can really make for an interesting trip.

10 Meter FM Repeaters

If your rig has FM capability, there's a world of DX with near-studio-quality audio awaiting you on the 10 meter FM repeater band. (Propagation permitting, of course.)

Although there are only four repeater outputs in the band plan (29.62, 29.64, 29.66 and 29.68 MHz), there are more than 100 repeaters listed in the current *ARRL Repeater Directory*. Using these machines is simple.

Because a 100-kHz *offset* is employed, you'll need to program your rig to transmit 100 kHz below the output frequency of the repeater you want to work. (Various rigs accomplish this split frequency operation in different ways. Check your rig's instruction manual for setup information.) Once the rig is ready, operation is much the same as on VHF/UHF repeaters. Just key up and give a call. But don't be surprised if you hear hams answering you on several different repeaters. It's a common occurrence when the band is open. Usually, FM's capture effect will permit you to hear only the repeater with the strongest signal, but rapidly changing band conditions can shuffle signals in and out like a deck of playing cards.

Nonetheless, 10-meter repeaters can be a lot of fun. Give them a try, and remember—since FM operation is a 100% duty cycle mode, you'll need to reduce your rig's RF output to prevent overheating. Follow the instructions in your rig's manual.

CW: Digital Operation for your Digits

In my desk I have a QSL card from a ham I worked while I was CW mobile. His comments say, in part, "I really feel as though I have been led down a primrose path. I can picture in my mind a guy sitting in an RV, running CW while someone else drives. Maybe I'm a skeptic, but sending CW and driving at the same time seems a bit far-fetched. . ."

Okay, I'll admit it really bolsters the ego of the mobile CW operator to have the masses think he's practicing a dark and secret art, reserved for those who possess the nerve and manual dexterity of a brain surgeon, but 'taint so. If you operate CW from home, you can probably do it mobile. The only prerequisite is you be able to copy in your head, since you obviously can't write what you hear and drive at the same time. But don't worry about it; copying CW without writing is a skill easily developed with a bit of practice.

Many first-time mobile CW ops find making their first few contacts while stationary is a great confidence builder before going out on the road. Once mobile, making your initial contacts with the keyer speed set somewhat below your capabilities will give you a leisurely opportunity to develop your CW "wings."

Operating style is a matter of preference. Many mobile CW operators find tuning to a clear frequency and calling CQ is most productive.

For example: I've done some listening and find a nice clear spot at 7.035 MHz. First I send, QRL? to make sure the frequency isn't in use. If I get no reply, I make a 3×3 call, CQ CQ CQ DE WF4N WF4N WF4N/M K. (I could have sent the /M after my call each time, but it isn't necessary.) As soon as I finish calling, I turn on my rig's RIT and sweep about 2 kHz each side of my transmit frequency—a necessity to hear replies that might be outside the passband of my rig's narrow CW filter. This filter is a must-have for mobile CW. When I get an answer to my CQ, I engage the dial lock to ensure against unintended frequency hops.

No matter how you establish contacts, you'll find lots of hams anxious to talk to you and ask questions about your

Fig 6-4—Although the author of this book uses a console between his bucket seats to hold his paddles, Eddie Cary, WA4EGQ, chose the classic technique of a leg strap. (*Photo courtesy of WA4EGQ*)

mobile setup. (The #1 most often asked question: "Do you have your key strapped to your leg?")

Of course, if you'd really like to be the center of attention, drive to a really rare county and check in on the County Hunter's net on 14.056 MHz. And you'll also find there are lots of general interest nets on the CW subbands. Refer to *The ARRL Net Directory* for more information.

Although it may be challenging at first, mobile CW can quickly become a pleasurable obsession. So consider yourself warned. Some hams have been known to deviate many miles from their planned routes just to extend their operating time.

Doing It *Your* Way

Although our primary focus has been on mobile

operation from conventional passenger vehicles, we must not ignore the fact that hams, being the diverse group of people they are, find lots of other ways to go mobile.

Aeronautical Mobile: Taking Amateur Radio to New Heights

As I sat in my shack on a Saturday evening putting the finishing touches on the next day's Sunday School lesson, my 2-meter rig, tuned to 146.52 MHz simplex, suddenly crackled to life.

"Okay, the QTH is about 50 miles south of Nashville, Tennessee. There are several other stations in here so I won't hold it. 73. . . ." As I reached to turn down the volume just a bit, I was thinking aloud, "Hmmm, that guy's about 150 miles from here. Must be a pretty good band opening." As I redirected my attention to my notes, I could faintly hear the same ham, apparently working another contact. "Very good old man, nice to meet you and 73 from . . . mobile."

This got my attention. A VHF mobile 150 miles away? I sure wanted to know all about his radio setup. I cranked up the volume on the rig and soon he was back. With the skill of a seasoned contester, he was finishing another contact, his signal now becoming R-4. "Take care and 73 from KB9ISD, aeronautical mobile." "That explains it," I shouted, "the guy's in an airplane." I picked up the mike and gave him a call. Immediately he replied, "WF4N from KB9ISD. The name here is Tracey." As we chatted, Tracey told me he was on his way home to Indianapolis, having flown to Cancun, Mexico, earlier in the day. He was running about 2 watts from an ICOM 24AT handheld and rubber duck, which was doing a pretty good job from an elevation of 37,000 feet. Of course, he had the advantage of a window seat. He was flying the plane (an L-1011 with 362 passengers onboard).

In consideration of the other stations who were hearing

only Tracey's side of the QSO, I soon signed, commenting that his route would take him almost directly over my house. "It won't be long before I get there," he noted as he signed, "I'm coming your way at eight miles a minute. 73 from KB9ISD aeronautical mobile."

As Tracey answered another caller, I turned to see my 8-year-old daughter, Whitney, standing mesmerized behind me. "That guy is going to be flying over our house?" she asked. "Sure is," I replied in my best Joe Cool matter-of-fact voice. "Let's go outside and watch for him," she suggested. "Why not?" I replied, sticking my HT in my pocket as I went out the door.

Sure enough, in just a few minutes, Whitney spotted a flashing red light gliding along under the twinkling stars of a dark southern sky. As the light grew closer, I turned on my HT and heard Tracey signing with another station.

"KB9ISD from WF4N. Got time for one more?" I asked. "Sure Roger, go ahead," he replied. I explained what we were doing, and that Whitney was really enjoying watching him pass overhead as I chatted with him. Obviously entertained by the novelty of it all, he answered, "That's great. Tell her to look real close 'cause I'm waving at her right now."

In the remaining minutes before he began his descent into Indianapolis, Tracey explained he had upgraded from Novice to General only a couple of days earlier, and in addition passed his 20-wpm code test.

Of course you don't have to be the captain of a commercial airliner to operate aeronautical mobile. Hams have taken to the skies with their rigs in private planes, ultralights, gliders, hot air balloons, and of course, the space shuttle.

If you pilot a private craft, going aeronautical mobile is as easy as taking along your HT on your next flight. A brief call on a simplex frequency (146.52 MHz is a favorite for many pilots) announcing you are aeronautical mobile, is guaranteed to bring the kind of response normally reserved for rare DX.

Since the distance to the horizon is more than 60 miles when you are at an elevation of only 2000 feet (distance to the horizon increases as the square root of the increase in elevation), the line-of-sight limitation of 2 meters becomes nearly meaningless when you are aeronautical mobile. Consequently, you can make lots of contacts while running low power (2 watts is plenty), and using only a rubber duck antenna.

Intoxicated by such phenomenal range, you might be tempted to set a new distance record for working the club repeater, but consider the potential for interference. Hams in adjoining states will take an understandably dim view of being wiped out on their local repeater by a ham 200 miles away. (I once tried to call a friend on our club's repeater while I was flying at 3000 feet, nearly 100 miles away. When I unkeyed, the sound of numerous repeaters simultaneously identifying greeted me like a nest of angry hornets.)

Speaking of noise, if you've spent any time in smaller aircraft (especially prop-driven aircraft), you know what a problem noise can be. The use of an earphone or headset can be a tremendous aid to your operation but don't forget the people you work are hearing all that racket too. Perhaps you'll want to acquire a good noise canceling microphone for use in your craft.

Powered aircraft are arrayed with a variety of electronic communication and navigation equipment; the ramifications of having your ham gear interfere with it are obvious. If you are a private pilot, the operation of Amateur Radio gear by you or your passengers will obviously be at your discretion. Remember, an amateur station being operated on a plane flying under IFR (Instrument Flight Rules) must comply with all applicable FAA rules.

Ever wonder why you don't hear very many hams operating from commercial airliners? The reason is simple. Most airlines specifically prohibit the use of Amateur Radio gear aboard their flights. Only if the pilot's permission is sought and received will you be allowed to use your radio on board the plane. And this rule makes good sense. How would you like to see a 747 in a nose dive toward your backyard because a ham on board was making an autopatch with his HT?

Okay, getting permission to transmit is too much hassle, so you'll just take along the HT to listen. Right? Wrong. The local oscillator in your rig can also cause interference to sensitive electronic systems on a plane. Unless you've received permission to do otherwise, leave your radio gear off and stored away in your luggage. Violating FAA rules pertaining to the use of radio equipment on board a plane can land you in a lot of trouble or worse.

Up on Two Wheels

I remember my first motorcycle—a Honda 90. Now to an 11 year old, that was a big bike.

Have you taken a look at today's lavish touring bikes? Some of them resemble two-wheeled locomotives. With lots of room, and plenty of power (many riders tow trailers with their bikes), these big rigs are a natural for mobile hamming. If you want to read a truly intriguing story about motorcycle mobiling, check out the article by Al Brogdon, K3KMO, in July 1993 *QST* where he chronicles his 10,500 miles of motorcycle mobile CW operation to Alaska and back.

Operating a radio from a motorcycle presents some interesting though certainly not insurmountable challenges. Naturally, placement and mounting of rigs and antennas must not compromise your ability to safely operate the bike. Letting a wheel drop off the shoulder of the road with a car is one thing, doing it on a motorcycle is guaranteed to decrease your life expectancy. Some states have laws governing the placement of radio equipment on motorcycles, so check your state's motor vehicle safety codes before you mount your rig. Of course you'll also want to take security into consideration when installing a rig on a motorcycle.

Anything beyond the most casual operation begs for the incorporation of an audio system (for transmit and receive) into the driver's helmet. Most hams adapt one of the commonly available systems designed for CB use or driver-passenger intercoms. Use one with a VOX circuit, and you truly have "hands off" radio and "hands on" biking.

Pedal Power

The absence of a powerful engine and electrical system doesn't have to be a deterrent to two-wheeled mobile operation. Talking to your ham buddies while you pedal flat-out for several miles is good fun and great exercise.

Unless you are willing to endure the weight penalty of an HF rig and the accompanying storage battery (yes, it has been done), bicycle mobile operation is best reserved for the VHF/UHF bands. With an HT's limited power output, a good antenna will be critical to the successful operation of your bicycle mobile station. You'll reap the best results by tossing out the usual puny little rubber duck antenna and mounting a standard mobile antenna on the bike. Of course, you'll find the practically non-existent ground plane offered by a bicycle will make antenna tuning difficult or impossible. Fortunately, a suitable counterpoise is easily fabricated. You might want to refer to the March 1993 *QST* article "Bicycle-Mobile Antennas" by Steve Cerwin, WA5FRF and Eric Juhre, KØKJ/5 for detailed information about mounting and tuning bicycle mobile antennas.

Maritime Mobile: On the High Seas or Inland Lakes

Are you an accomplished sailor? Or maybe you are planning a vacation cruise on an ocean liner. Then why not complement your time on the water with some ham activity? Elaborate equipment isn't necessary. Water makes an excellent ground plane, and good performance is possible with only a modest antenna. Signing maritime mobile using a foreign call sign can easily put you at the receiving end of an enormous pileup.

How you operate maritime mobile is limited only by your imagination. VHF/UHF, HF, phone, CW, and the digital modes are all practical candidates. If you want to converse with other sailing hams, check out the maritime mobile service nets. You'll find the Waterway Radio & Cruising Club Net on 7.268 MHz, the Maritime Mobile Service Net on 14.300 MHz, the Sailfish Net on 28.425 MHz and the Ten Meter Maritime Mobile Net on 28.380 MHz. Consult *The ARRL Net Directory* for days, times and coverage areas of these nets.

If you operate from US waters on a craft registered in the US, the same FCC rules governing operation from your home station will apply. (I know a ham who takes great pleasure in signing maritime portable on 2 meters from his bathtub.) However, if you will be operating from a vessel registered in another country, or from the territorial waters of another country, the rules change.

If you operate in US territorial or international waters from a US-registered vessel, the FCC rules applicable to your class

Fig 6-5—Thomas "T. Bone" Whatley, KI4FD, uses this solar-powered rig for mobile as well as Field Day operation. Two golf cart batteries back up the solar panel on cloudy days. A "Carolina Bug Catcher" antenna on the front left of the canopy and an electric trolling motor at the stern complete the setup. Note the fossil-fuel lantern for night-time operation. (*Photo courtesy of KI4FD*)

of license will apply. If you are operating from international waters aboard a vessel of foreign registry, you must acquire a license or permit from the ship's country of registry. If the ship enters the territorial waters of a foreign country, you'll be responsible for knowing and observing the regulations of that country in addition to those of the ship's country of registry.

Sounds complicated, but it really isn't. Consult *The ARRL FCC Rule Book*, or contact the Regulatory Information Branch at ARRL HQ before you set sail.

Keep It Legal

In addition to the usual FCC regulations, when we take our radios out on the road we come under the jurisdiction of additional laws that can affect how and when we operate. Before you go mobile, determine what laws apply to your operation.

You might score big points with your non-ham passengers by using an earphone or headset, but make sure the device you use leaves you with one ear available to monitor for warning sounds from other vehicles. Not only is covering both ears with headphones or a headset very dangerous, in some states it is very illegal.

If you will be using a VHF/UHF rig, capable of receiving public service frequencies (such as police and fire), be careful you don't run afoul of a "scanner law." Intended to prevent mobile criminals from using scanning receivers to monitor police activities, scanner laws have been an unwelcome (and sometimes unexpected) constraint for many hams. Although most states exempt licensed Amateur Radio operators from their scanner laws, and the FCC has gone on record as supporting a nationwide exemption, make sure you know the particular laws for the state or states where you'll be operating. In addition, it is a good idea to know the revised statute number for the applicable law. It's pretty frustrating to be detained for a long period of time while a policeman checks to determine the existence of a law you claim exists, but of which he isn't aware. It also may help to have a copy of your amateur license with you, to prove you are a ham.

Some radios with extended receive coverage present an additional hazard because they are capable of transmitting outside the amateur bands. Why is this a hazard? Amateur Radio gear doesn't carry FCC Type Acceptance (approval) for transmitting outside the ham bands, so it is illegal to use it for this purpose.

Okay, so you have no plans to ever use your "opened

up" rig to transmit anywhere but the ham bands. But what if you do it accidentally? Many mobile hams have experienced the embarrassment of sitting on the mike, allowing anyone monitoring the frequency to hear what (or who) they talked about when they thought no one was listening. Having your local police department hear your "Candid Camera" conversations could easily qualify as the ultimate embarrassment and more. Don't jeopardize your license. If you are contemplating the mobile use of a modified rig, you may want to consider reversing the modification that allowed out-of-band transmit. Or, when you program your rig with the public service frequencies you plan to monitor, program in a transmit frequency inside the amateur bands. Then when your mike key sticks for a half hour as you drive down the freeway, your pride will be the only thing that suffers.

Reporting an Emergency

One of the best reasons in the world to have a mobile station is for emergency communication. If you should happen upon the scene of an accident, you might find you are the only person there with the means to summon emergency assistance. In situations involving serious injuries, time is of the essence. Your first priority is to get EMS on the way immediately.

As soon as you determine injuries are involved, establish your radio connection with EMS dispatch. If you are using a familiar repeater, state you are making an emergency autopatch and dial the access code and emergency number.

If you don't know how to access the autopatch, give your call and announce you have emergency traffic and need immediate assistance. Someone can then bring up the patch for you or make the call from their home phone. If the repeater is in use, don't be shy. You have priority on the frequency.

If HF is all you are able to use, then use it. In most cases it will be quicker than driving somewhere to a phone.

When the call is answered, tell the dispatcher you are reporting an accident with injuries. Speak in short sentences, and release the push-to-talk button between sentences to allow the dispatcher to break in with any question they may have. Give them the opportunity to tell you they already have the information, rather than tying up their emergency line with a three minute, fact filled monologue.

You can tell them you are a ham, but most EMS personnel I spoke with stated that it isn't really necessary. Some dispatchers will request your name before hanging up. Give the location of the accident. Include the highway number, nearest mile marker or exit number if you know it, and the direction of travel if you are on a divided highway. Be absolutely certain of your information. Routing the EMS to the wrong location is a potentially tragic mistake you don't want to make.

If you know the number of victims, give this information to the dispatcher. Tell them if there is a hazardous substance involved, any fuel leakage or a fire. Don't waste valuable time or risk causing confusion by including non-essential information, such as your call or the repeater ID. Stay on the line until the dispatcher indicates it is okay to hang up. If possible, remain at the scene to render whatever assistance you are qualified to give and to provide additional communications if necessary.

CHAPTER 7

Dealing With Automotive Interference

MI, RFI. These are ingredients in the alphabet soup of interference being served up to many mobile Amateur Radio installations. This intoxicating witches' brew can impart some very strange side effects on your auto, your rig, or both. The symptoms can include windshield wipers that operate when they aren't supposed to, (or don't operate when they are supposed to), engines sputtering, spitting or whispering mechanical obscenities, erratic operation of electronic displays—or conversely, popping, crackling, whining or whistling noises in your rig's audio (during transmit, receive or both). The list goes on!

What is EMI?

Since the dawn of Amateur Radio, hams have wrestled with radio frequency interference or RFI. At first, the noise generators were simple: A toaster, a furnace, perhaps a pole-mounted power transformer. Then came the electronic age. Suddenly, it seemed as though any device that relied on the flow of electrons to function was capable of causing and/or being affected by radio interference. Along with all the new sources of RFI came a new term, Electromagnetic Interference, or EMI. What's the difference between RFI and EMI? Simply, RFI is EMI, but EMI isn't necessarily RFI. The broader term addresses interference, regardless of frequency or the means of transmission—it might be radiated interference, arriving at the rig via the antenna, or conducted interference, traveling along wires or cables.

Three principal players team up to generate EMI.

First, there must be a source of interfering signal. This can be a radio transmitter, an electric motor, or perhaps something as basic as an electrical switch.

Second, there must be a recipient of the interfering signal—the affected device. This can be your rig, or one of your auto's electronic systems.

Third, the interfering signal must have a means of transmission to the affected device. Wires, cables and open space are all good candidates.

Hmmm. Sounds a lot like what's required for ham radio communications, doesn't it? In fact, if you've ever had a QSO clobbered by someone tuning up on the frequency you were using, you've experienced one of the most basic forms of EMI.

Caveats

In this chapter we'll look at some of the EMI problems faced by the mobile operator. We'll show you how to track down interference—and show you ways to cure it. Keep in mind your dealer is the best source of information about your particular vehicle. All automakers provide their dealers with Technical Service Bulletins (TSBs), some pertaining to radio/auto EMI problems. If your dealer can't or won't assist you, don't give up. Contact the manufacturer directly. To learn how the car manufacturers deal with EMI problems, refer to the September 1994 *QST* article, "Automotive Interference Problems: What the Manufacturers Say," by Ed Hare, KA1CV.

Unless you have experience with automotive electronic systems, it's best to have any EMI-solving modifications performed by a qualified mechanic/technician. This is especially true if your vehicle is under warranty, since owner-performed modifications can void the warranty, leaving you responsible for expensive repairs. Remember, improperly performed modifications can dangerously affect critical braking, engine or transmission control systems.

Even if you won't be performing any EMI fixes on your vehicle, you can save your mechanic some valuable time and avoid having ineffective work done, by pinpointing the source of interference when you take your vehicle to the shop.

Proceed carefully with your troubleshooting. While tracking down a source of interference, you may find it necessary to temporarily remove power from a circuit or system while the engine is running. This is best accomplished by pulling the associated fuse from the fuse panel. Only if it's absolutely necessary should you disconnect the wiring harness plug from an energized electronic module, and then only after consulting a service manual or your dealer's service department.

Be aware some vehicles will reward your fuse-pulling efforts by illuminating the "check engine" light and placing the engine control system in the power-limited "limp home" mode. If this happens, don't panic—it doesn't necessarily mean you've done any damage. It's just that the affected control module has sensed a failure in a monitored system and has taken measures both to prevent drivetrain damage and to alert the driver of the problem. If this happens, shut the engine off and disconnect the positive battery cable for about five minutes. This will clear the fault code from the memory of your car's computer and restore normal operation of your vehicle. (It probably will also require you to reprogram the radio presets and reset the clock—sorry.)

The approach you take to curing EMI problems will often be affected by how the noise arrives at your rig—either through the antenna and feed line, or the power leads. How do you determine which one is the culprit? Easy! Disconnect the feed line from your rig. Does the noise remain? Then it's coming to the rig via the power leads. (Make sure you don't have other leads for keyers, remote speakers, etc, connected during this test.)

If the noise is only present with the antenna connected, then it is being picked up by the antenna. To be sure, leave the feed line connected at the rig, but disconnect it at the antenna. If the noise level remains much the same, then your feed line is functioning as an EMI antenna. Rerouting the feed line away from the source of interference, or switching to cable with better shielding will usually help.

One of the things that make FM radio so popular with hams is its immunity to noise. The detector in an FM receiver doesn't respond to amplitude modulation (AM), and most broadband noise is predominately AM. Even so, your FM rig can be experiencing interference without you being aware of it. Although you don't hear it in the audio, AM noise can partially or completely mask received signals. This type of interference is easy to spot. It usually causes an S-meter reading even though the squelch remains closed. In less severe cases, it may ride in on top of the audio from a received signal, making it unreadable.

With the surging proliferation of electronic devices in today's automobiles, there are literally dozens of potential EMI generators poised, ready to trash your QSOs. Let's look at some of the villains.

King Spark Rides Again

The ignition system is perhaps the most common source of interference found in an automobile. Supplying upwards of 50 kV to each spark plug, it's the ultimate spark-gap transmitter. Ignition noise is readily identified as a rhythmic popping or ticking in the receiver audio. It varies directly with engine speed, the amplitude becoming greater as engine loading is increased.

Normally, the first strike in the assault on ignition noise is to install resistor spark plugs and wires. Now, however, most if not all vehicles come with resistor plugs and wires as standard equipment. Often, installing new spark plugs and wires will help to suppress ignition noise, since many

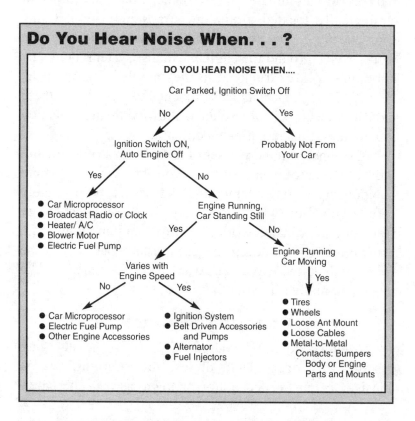

Do You Hear Noise When. . . ?

DO YOU HEAR NOISE WHEN....

Car Parked, Ignition Switch Off

No — Ignition Switch ON, Auto Engine Off

Yes — Probably Not From Your Car

Yes:
- Car Microprocessor
- Broadcast Radio or Clock
- Heater/ A/C
- Blower Motor
- Electric Fuel Pump

No — Engine Running, Car Standing Still

Yes — Varies with Engine Speed

No — Engine Running Car Moving

No:
- Car Microprocessor
- Electric Fuel Pump
- Other Engine Accessories

Yes:
- Ignition System
- Belt Driven Accessories and Pumps
- Alternator
- Fuel Injectors

Yes:
- Tires
- Wheels
- Loose Ant Mount
- Loose Cables
- Metal-to-Metal Contacts: Bumpers Body or Engine Parts and Mounts

ignition-related EMI problems are caused by faulty ignition system components. With the extremely high energy levels produced by today's ignition systems, most vehicles will continue to perform somewhat normally even though some components have deteriorated to the point that they are producing excessive interference. The spark plug's porcelain insulator can crack. The plug wire's resistive core can break, and the insulating jacket can be broken, cut or burned. Any of these conditions can cause noise-producing arcs of high voltage. In addition to checking spark plugs and wires, distributor caps and rotors should be inspected for corrosion, cracks and carbon tracking (an indication of arcing).

If a tune-up of the ignition system doesn't sufficiently reduce or eliminate interference, the addition of strategically located grounding straps may help. Using short, heavy copper straps (shield braid from RG-8 cable works well), ground the engine, exhaust system, bumpers, hood and deck or trunk lid to the chassis or body of the vehicle. Though mechanically attached to the vehicle, these parts may be electrically isolated, allowing them to function as antennas, thus radiating electrical noise.

Extremely stubborn cases of ignition noise may require the shielding of some or most of the ignition system components. Before attempting this measure, be absolutely certain you know what you are doing. Misdirected energy from an unterminated spark plug wire can jump to ground, using other wires, or even rubber hoses as a pathway. If the spark finds its way to an electronic sensor or module, the result can really ruin your day.

Charging Systems

Another potential source of EMI is the charging system. In modern automobiles, the charging system consists of an ac generator, known as an alternator, controlled by a solid state voltage regulator. Because the ac is rectified, but not filtered, within the alternator, the output can contain ripple, just like a poorly filtered power supply.

Charging system noise is conducted through the vehicle wiring to your rig, where it affects the audio sections of the transmitter and receiver. Alternator noise is easily identifiable as a whine in received, or more noticeably, transmitted audio. Temporarily disabling the charging system can further identify alternator-produced interference. The pitch as well as the intensity of alternator whine can vary with engine speed and the loading of the charging system. That's why some installations experience a greater level of alternator whine at night, when vehicle lights are on.

If your mobile rig is plagued with charging system noise, first check for loose or corroded connections in the wiring between the alternator and the battery. Since the alternator's pulsating dc output is filtered by the battery, even a small amount of resistance can affect noise levels. So can a malfunctioning voltage regulator—check for correct charging system output voltage. The car shop manual usually will give you the conditions for this test.

If noise persists with system components in good condition, adding filters may help. Alternators usually have an internal capacitor connected from the output to ground, to help suppress noise. Connecting an additional, externally mounted capacitor from the output to ground can further reduce noise levels. The value of the capacitor isn't critical—0.47 µF is usually satisfactory for HF and 0.01 µF for VHF. You may also wish to install a filter in your rig's power leads, located as close as possible to the radio. Radio Shack stores, and others, offer various types of dc filters— be sure to choose one that can handle the current required by your rig.

Older Vehicles

What's the worst thing about a new automobile? It doesn't stay new! Time and miles take their toll on the best of automobiles. Paint fades, chrome rusts, and eventually the vehicle develops the familiar "used car" look. At the same time, weather, vibration and road salt have been working on the unseen areas of your car. As a result, older vehicles sometimes become very prolific generators of electrical noise as various body and chassis mating surfaces rust and their fasteners loosen. Tightening loose body bolts, cleaning and repainting rusty joints or bonding across joints with ground straps can sometimes lessen interference. Try adding ground straps from various points of the exhaust system to the car frame.

Worn-out or broken engine mounts are another potential source of interference, allowing intermittent metal-to-metal contact of the mount components, which are normally kept apart by an isolating filler made of rubber. Cure EMI with new engine mounts? It's possible.

Electric Motors

Automobiles usually have several dc motors, most all capable of generating EMI well into the VHF region. They include windshield wipers, heat/air conditioning fans, power windows and seats, throttle body idle controls and electric fuel pumps. Because of the variety of different motors used in an automobile, it's difficult to know what type noise to expect. Generally, dc motor EMI will manifest itself as a frying or crackling sound, or sometimes as a whine similar to that generated by the charging system.

Recognizing dc motor noise is one thing—identifying the source can be a horse of an entirely different color. Obviously, if the culprit is a manually controlled motor (power seat, power window, heater fan, etc), you can use the process of elimination to determine the specific motor causing the interference. Unfortunately, most vehicles also employ a number of motors that operate automatically, or are actuated by an electronic controller (idle speed controls and electric fuel pumps are two of many possibilities). Tracking down the guilty suspect in this group can be challenging, to say the least.

Possibly the best method, although it may be expensive, is to have your mechanic use his diagnostic analyzer's actuator test mode to operate each suspect motor while you monitor your radio for interference. If your vehicle's control systems don't allow an actuator test to be performed, try disabling suspect motors one at a time by pulling the appropriate fuse, power lead or control relay. Again check your vehicle's service manual or your dealer's service department before pulling plugs.

Don't be surprised if your search reveals the electric fuel pump as the source of your EMI woes. It's one of the most commonly reported causes of EMI. Due to its location in or near the fuel tank, its long power leads provide an excellent antenna for the radiation of EMI. As with other dc motors, fuel pump EMI can have various characteristics. Identification is easily accomplished by simply pulling the fuel pump fuse from the fuse panel while the engine is running. (The engine will stall if the fuse is not quickly replaced.)

Because dc motor EMI can be conducted or radiated to your radio, stamping it out requires you apply the cure as close to the source (the motor) as possible. If the problem motor is a common cause of radio interference, it's possible the vehicle manufacturer makes an EMI filter available for that particular application—check with your dealer.

If the manufacturer doesn't offer a filter, you may wish to try one of the commercially available noise filters. Or if you are a do-it-yourselfer, you may want to construct your own. Either way, place the filter in the power leads as near the motor as possible, and make sure the filter is rated to handle the current drawn by the motor. **It is best to have a qualified service shop do all work near gas lines and the gas tank.**

Those Marvelous Microprocessors

From shuttle launches to microwaved lunches, computers have immeasurably affected the way we live. Just about every device using electricity is now controlled by a computer, or its more basic building block, the microprocessor. Most automobiles employ several microprocessors, controlling a wide variety of functions such as fuel, spark, or braking systems, even the clock. Without question, the marriage of the microprocessor and the automobile has helped create a level of automotive sophistication we all have grown to know and appreciate. Unfortunately, as many mobile hams have discovered, it isn't always a match made in heaven.

Microprocessors require a clock signal to function. This signal is generated by an oscillator circuit, its frequency controlled by a crystal or ceramic resonator. Although this low power mini-transmitter may be operating at a frequency far below the amateur bands, the harmonics from its square wave output can extend well into the VHF region. Consequently, your mobile rig can be plagued with spurious signals occurring at regularly spaced frequency intervals, or by broadband digital noise extending through an entire band or bands.

Curing microprocessor EMI requires you first identify and locate the particular electronic control module (ECM) that is producing the interference—your vehicle's service manual will be an indispensable aid to this task. Using a receiver that tunes the affected frequency or frequencies, establish under what conditions the interference occurs (engine running or not, for example). If you are using an HT, you may be able to locate the offending ECM by slowly sweeping the vehicle with the HT while you observe for a peak in signal strength. Pull fuses at the fuse panel one at a time, until the interference ceases. Some fuses may power more than one module—often true of engine control modules. Some modules also receive power from other modules—which can further frustrate even the best of EMI sleuths. In those cases, it will be necessary to disconnect the power carefully at each individual module. If you suspect an engine control module, first try disconnecting power to the module with the ignition *on*, but with the engine *not* running. Usually, noise from engine control modules will be present anytime the ignition switch is on, regardless of whether or not the engine is running. It is usually not a good idea to disconnect an engine control module when the engine is running.

If you find an ECM is the source of interference, check with your dealer to see if an EMI suppressed replacement is available—possibly on a no-cost exchange basis.

Many noisy ECMs will benefit from additional or improved shielding. If the ECM is in a non-metallic enclosure, you might try fabricating a shield from light gauge aluminum or copper (or copper tape). Be sure the shield is well grounded to the vehicle chassis at only one point, and be careful not to defeat cooling slots or vents in the ECM case.

You might find some metal ECM enclosures are not grounded at all, having been mounted on rubber shock-isolating hardware. A short, wide ground strap from the case to the vehicle chassis will usually provide a marked reduction in EMI—it may be necessary to experiment to determine the most effective point of connection for the ground strap.

Since much or most of ECM interference can be radiated or conducted via the vehicle wiring harness, often you will need to apply suppression measures to the harness where it attaches to the module. Obviously, with dozens of conductors emanating from many ECMs, filtering of individual wires is not practical. Instead, use ferrite toroids or snap-together ferrite cores as chokes to block transmission of EMI into the wiring harness. Place snap-together cores over the harness, as close as practical to the ECM. If there is sufficient space and harness length available, you can wind the harness around a large toroid for the same effect. Consult the core manufacturer's catalog to determine the core's effectiveness in suppressing noise at the frequency of interest. You may want to tape the core and wire together and then to a nearby bracket. The weight of some cores and continuous flexing can damage the wire insulation.

Finding RF in all the Wrong Places

Having your rig interfered with by automotive electronics is disruptive to enjoyable hamming, but it's a much more serious problem when your transmitter interferes with your vehicle's electronics. Those undesired vehicular reactions (known as susceptibility) can be as minor as erratic readings from instruments or gauges, or as major as having the engine stall every time you key your rig.

Although the previously described methods of suppressing ECM-generated EMI are also effective in reducing susceptibility, it's a good idea to first inspect your installation for any shortcomings that might be causing problems.

Make sure power and antenna cables are routed as far

away as possible from vehicle wiring and electronic systems. Properly mount the antenna, preferably on the roof or the extreme rear of the vehicle. Tune the antenna system for the lowest possible SWR. Check the coaxial cable, and see if it is properly grounded. The shield coverage should be at least 95%. In many years of mobile operation from several vehicles, I've had two run-ins with the susceptibility bug—both caused by feed line ground problems.

Try Before You Buy

Are you planning the purchase of a new vehicle to be the future home to your radio equipment? If so, a pre-purchase investigation of potential electronic systems/radio equipment compatibility problems is a wise move.

Your first step should be to check with the dealer's service department to see if there are any technical bulletins pertaining to radio installations in the vehicle. Some dealers may be unaware of the existence of those service bulletins—it may be necessary to contact the manufacturer directly.

If you plan to order a new vehicle, check to see if it is available with optional EMI suppression equipment. As an example, the options included on fleet vehicles should be considered. They are specially equipped for police, fire or taxi service, where two-way radio installations are common.

If you are considering a vehicle in a dealer's existing inventory, it's a very prudent idea to test for EMI problems before you buy. Since EMI problems are a two-way proposition, you'll want to check for both vehicle-to-radio interference and radio-to-vehicle interference.

Run Your Own EMI Test

Interference to your radio equipment is by far the

easiest to check. All you need is a battery powered receiver tuning the frequencies you plan to operate, and an antenna. Although it's possible to use a whip or rubber duck antenna for your tests, a mag-mount antenna will give better results and allow you to try different antenna mounting locations as you observe the effect on received interference. Take along a polyethylene sandwich bag or something similar to place under the magnet to prevent damage to the car's finish. Select an antenna with a well-shielded feed line for this test.

With the receiver on, start the vehicle engine. As you tune the receiver through its range, observe for broadband noise such as ignition or dc motor noise, as well as modulated or unmodulated spurious signals. If possible, use the receiver's AM mode since most broadband noise is amplitude modulated and won't be detected by an FM receiver. Lacking AM capability, observe the receiver S-meter for the presence of AM noise.

Be sure to operate all power accessories (heater, wipers, even the lights) individually, and in combination. If there is a cellular phone installed, turn it on and see what, if anything, happens.

Have an assistant drive the vehicle for a few miles as you observe for problems that may be present only when the vehicle is in motion. Since many of the engine's operating conditions change as it warms up, be sure to allow it to reach normal operating temperature before you finish the test.

What if your test reveals some EMI problems? Experiment with different mounting locations for the antenna to see if there is one that eliminates, or at least diminishes the noise. If this measure is unsuccessful, you'll likely have to decide if the interference is severe enough to warrant shopping for a different make or model of vehicle.

If you observed only low or moderate levels of broadband noise, there's a good chance a permanent installation will benefit from filtering. If the spurious signals weren't on frequencies you expect to be using, they probably won't be a great annoyance.

Will My Radio Zap My Car's Computer?

Having to cope with interference from vehicle electronics can be an irritating, yet tolerable situation for the mobile ham. But how do you know the vehicle will be equally as tolerant of interference from your radio? Obviously, it's best to test the vehicle for susceptibility to interference from your transmitter—before you buy.

Since you will have to transmit with a rig (temporarily) installed in the vehicle, your first step will be to obtain permission from the dealer to perform the test. Don't be surprised if he is reluctant or even refuses to allow such a test. After all, he's a businessperson, not an electrical engineer. Keep in mind, if the dealer isn't willing to cooperate in order to make a sale, he will probably be even less cooperative if you buy the vehicle and experience EMI problems later on. For your own protection, if permission is granted, ask the service manager or his appointee to be present to observe the test.

To test the candidate vehicle, you'll need a rig (or rigs), a mag-mount antenna and a battery capable of powering your rig to full output. Although the test might be more revealing if you power the rig from the vehicle battery, permission to test will be much easier to obtain if there will be no electrical connections made to the vehicle.

With the engine running, begin by making short transmissions on each band of interest with the transmitter adjusted for minimum output. Repeat with gradually increasing power levels until all bands have been checked at full power. Perform the test with as many accessories (radio, heat/air conditioning system, etc) operating as possible. Of course, it goes without saying, to prevent damage you should stop transmitting at the first indication of an adverse reaction from the vehicle.

You might get some interesting information by parking your auto next to the new car and transmitting from your current installation at high power. You might even open the hood of the new car while running the test.

If your setup and the dealer allow, include an on-the-road test to check for EMI effects on transmission, braking and cruise control systems. Be sure to perform the tests in an area where a sudden reaction of the vehicle won't pose a hazard to other motorists—a large, vacant parking lot perhaps.

The Bottom Line

Okay, so the transmitter test revealed some susceptibility problems, and trying various locations for the mag-mount antenna didn't alleviate them. What now? Since both the dealer and manufacturer are in the business of selling cars, hopefully they will make an effort to cure the problem. If not, and the interference is minor, you may opt to either live with it or attempt a solution of your own. Be sure to inquire what effect your modifications might have on the warranty.

If you should find yourself faced with one of the extremely rare instances where critical vehicle systems are affected, you'll have to either choose another vehicle or abstain from mobile operation. (That's a choice?) Believe it or not, there are a few automakers that don't sanction the use of radio transmitting equipment in their vehicles.

Where Do We Go from Here?

Automotive EMI problems aren't new. As long ago as

the late '60s, hams were reporting their transmitters were having strange effects on their automobiles. Even so, it's doubtful anyone could have envisioned the automobile of the future would present such a potential EMI nightmare.

Although it's probably fair to say most automakers are in a catch-up mode when it comes to dealing with EMI problems, take comfort in the fact that it is an issue that is receiving an ever-increasing amount of attention. In fact, if you've attended the Dayton HamVention recently, you may have been surprised to find representatives of two major automakers there to ask and answer questions pertaining to EMI.

With the explosive growth of the mobile communications industry, incompatibility between automotive electronics and communications equipment is an ongoing problem. In time, it may well become only a relic of "the old days." A good reference on curing EMI problems is *Radio Frequency Interference: How to Find It and Fix It*, published by the ARRL. A separate chapter covers mobile EMI problems and solutions.

JOIN ARRL TODAY AND RECEIVE A *FREE* BOOK!

I want to join ARRL. Send me the FREE book I have selected (choose one):

☐ *Repeater Directory*—gives you listings of more than 20,000 voice and digital repeaters throughout the US. ($7 value)

☐ *Your VHF Companion*—lets you explore the fascinating activities on the VHF bands: FM, repeaters, packet, CW, SSB, satellites, amateur television, and more. ($8 value)

☐ New Member   ☐ Previous Member   ☐ Renewal

_____
Call Sign (if any)   Class of License   Date of Birth

_____
Name

_____
Address

_____
City, State ZIP

Telephone Day (   )_____ Night (   )_____

Dues are $31 in US/$44 elsewhere (US funds). You do not need an Amateur Radio license to join. Individuals who are age 65 or over, upon submitting one-time proof of age, may request the dues rate of $25 in the US/$38 elsewhere (US funds). Immediate relatives of a member who receives *QST*, and reside at the same address may request family membership at $5 per year. Blind individuals may join without *QST* for $5 per year. If you are 21 or younger and a licensed amateur, a special rate may apply. Write to ARRL for details.

DUES ARE SUBJECT TO CHANGE WITHOUT NOTICE.

Payment Enclosed ☐

Charge to MC, VISA, AMEX, Discover No. _____

Expiration Date _____

Cardholder Name _____

Cardholder Signature _____

If you do not wish your name and address made available for non-ARRL related mailings, please check this box ☐.

THE AMERICAN RADIO RELAY LEAGUE, INC
225 MAIN STREET NEWINGTON, CONNECTICUT 06111 USA
(860) 594-0200   YMC 11/95
New Hams call (800) 326-3942

About The American Radio Relay League

The seed for Amateur Radio was planted in the 1890s, when Guglielmo Marconi began his experiments in wireless telegraphy. Soon he was joined by dozens, then hundreds, of others who were enthusiastic about sending and receiving messages through the air—some with a commercial interest, but others solely out of a love for this new communications medium. The United States government began licensing Amateur Radio operators in 1912.

By 1914, there were thousands of Amateur Radio operators—hams—in the United States. Hiram Percy Maxim, a leading Hartford, Connecticut, inventor and industrialist saw the need for an organization to band together this fledgling group of radio experimenters. In May 1914 he founded the American Radio Relay League (ARRL) to meet that need.

Today ARRL, with more than 170,000 members, is the largest organization of radio amateurs in the United States. The League is a not-for-profit organization that:
- promotes interest in Amateur Radio communications and experimentation
- represents US radio amateurs in legislative matters, and
- maintains fraternalism and a high standard of conduct among Amateur Radio operators.

At League headquarters in the Hartford suburb of Newington, the staff helps serve the needs of members. ARRL is also International Secretariat for the International Amateur Radio Union, which is made up of similar societies in more than 100 countries around the world.

ARRL publishes the monthly journal *QST*, as well as newsletters and many publications covering all aspects of Amateur Radio. Its headquarters station, W1AW, transmits bulletins of interest to radio amateurs and Morse code practice sessions. The League also coordinates an extensive field organization, which includes volunteers who provide technical information for radio amateurs and public-service activities. ARRL also represents US amateurs with the Federal Communications Commission and other government agencies in the US and abroad.

Membership in ARRL means much more than receiving *QST* each month. In addition to the services already described, ARRL offers membership services on a personal level, such as the ARRL Volunteer Examiner Coordinator Program and a QSL bureau.

Full ARRL membership (available only to licensed radio amateurs) gives you a voice in how the affairs of the organization are governed. League policy is set by a Board of Directors (one from each of 15 Divisions). Each year, half of the ARRL Board of Directors stands for election by the full members they represent. The day-to-day operation of ARRL HQ is managed by an Executive Vice President and a Chief Financial Officer.

No matter what aspect of Amateur Radio attracts you, ARRL membership is relevant and important. There would be no Amateur Radio as we know it today were it not for the ARRL. We would be happy to welcome you as a member! (An Amateur Radio license is not required for Associate Membership.) For more information about ARRL and answers to any questions you may have about Amateur Radio, write or call:

ARRL Educational Activities Dept
225 Main Street
Newington CT 06111-1494
(860) 594-0200
Prospective new amateurs call:
800-32-NEW HAM (800-326-3942)

W1AW schedule

Pacific	Mtn	Cent	East	Sun	Mon	Tue	Wed	Thu	Fri	Sat
6 am	7 am	8 am	9 am			Fast Code	Slow Code	Fast Code	Slow Code	
7 am	8 am	9 am	10 am			Code Bulletin				
8 am	9 am	10 am	11 am			Teleprinter Bulletin				
9 am	10 am	11 am	noon							
10 am	11 am	noon	1 pm			**Visiting Operator Time**				
11 am	noon	1 pm	2 pm							
noon	1 pm	2 pm	3 pm							
1 pm	2 pm	3 pm	4 pm	Slow Code	Fast Code	Slow Code	Fast Code	Slow Code	Fast Code	Slow Code
2 pm	3 pm	4 pm	5 pm	Code Bulletin						
3 pm	4 pm	5 pm	6 pm	Teleprinter Bulletin						
4 pm	5 pm	6 pm	7 pm	Fast Code	Slow Code	Fast Code	Slow Code	Fast Code	Slow Code	Fast Code
5 pm	6 pm	7 pm	8 pm	Code Bulletin						
6 pm	7 pm	8 pm	9 pm	Teleprinter Bulletin						
6^{45} pm	7^{45} pm	8^{45} pm	9^{45} pm	Voice Bulletin						
7 pm	8 pm	9 pm	10 pm	Slow Code	Fast Code	Slow Code	Fast Code	Slow Code	Fast Code	Slow Code
8 pm	9 pm	10 pm	11 pm	Code Bulletin						
9 pm	10 pm	11 pm	Mdnte	Teleprinter Bulletin						
9^{45} pm	10^{45} pm	11^{45} pm	12^{45} am	Voice Bulletin						

W1AW's schedule is at the same local time throughout the year. The schedule according to your local time will change if your local time does not have seasonal adjustments that are made at the same time as North American time changes between standard time and daylight time. From the first Sunday in April to the last Sunday in October, UTC = Eastern Time + 4 hours. For the rest of the year, UTC = Eastern Time + 5 hours.

• Morse code transmissions:
Frequencies are 1.818, 3.5815, 7.0475, 14.0475, 18.0975, 21.0675, 28.0675 and 147.555 MHz.
Slow Code = practice sent at 5, 7^1/$_2$, 10, 13 and 15 wpm.
Fast Code = practice sent at 35, 30, 25, 20, 15, 13 and 10 wpm.
Code practice text is from the pages of *QST*. The source is given at the beginning of each practice session and alternate speeds within each session. For example, "Text is from July 1992 *QST*, pages 9 and 81," indicates that the plain text is from the article

on page 9 and mixed number/letter groups are from page 81. Code bulletins are sent at 18 wpm. W1AW qualifying runs are sent on the same frequencies as the Morse code transmissions. West Coast qualifying runs are transmitted on approximately 3.590 MHz by W6OWP, with W6ZRJ and AB6YR as alternates. At the beginning of each code practice session, the schedule for the next qualifying run is presented. Underline one minute of the highest speed you copied, certify that your copy was made without aid, and send it to ARRL for grading. Please include your name, call sign (if any) and complete mailing address. Send a 9×12-inch SASE for a certificate, or a business-size SASE for an endorsement.

• Teleprinter transmissions:

Frequencies are 3.625, 7.095, 14.095, 18.1025, 21.095, 28.095 and 147.555 MHz.

Bulletins are sent at 45.45-baud Baudot and 100-baud AMTOR, FEC Mode B. 110-baud ASCII will be sent only as time allows. On Tuesdays and Saturdays at 6:30 PM Eastern Time, Keplerian elements for many amateur satellites are sent on the regular teleprinter frequencies.

• Voice transmissions:

Frequencies are 1.855, 3.99, 7.29, 14.29, 18.16, 21.39, 28.59 and 147.555 MHz.

• Miscellanea:

On Fridays, UTC, a DX bulletin replaces the regular bulletins. W1AW is open to visitors during normal operating hours: from 1 PM until 1 AM on Mondays, 9 AM until 1 AM Tuesday through Friday, from 1 PM to 1 AM on Saturdays, and from 3:30 PM to 1 AM on Sundays. FCC licensed amateurs may operate the station from 1 to 4 PM Monday through Saturday. Be sure to bring your current FCC amateur license or a photocopy.

In a communication emergency, monitor W1AW for special bulletins as follows: voice on the hour, teleprinter at 15 minutes past the hour, and CW on the half hour.

Headquarters and W1AW are closed on New Year's Day, President's Day, Good Friday, Memorial Day, Independence Day, Labor Day, Thanksgiving and the following Friday, and Christmas Day. On the first Thursday of September, Headquarters and W1AW will be closed during the afternoon.

Index

B

C

Please use this form to give us your comments on this book and what you'd like to see in future editions.

Where did you purchase this book?

☐ From ARRL directly   ☐ From an ARRL dealer

Is there a dealer who carries ARRL publications within:

☐ 5 miles   ☐ 15 miles   ☐ 30 miles   of your location?   ☐ Not sure.

License class:

☐ Novice   ☐ Technician   ☐ Technician with HF privileges
☐ General   ☐ Advanced   ☐ Extra

Name	ARRL member? ☐ Yes ☐ No
	Call sign _____

Daytime Phone (   ) _____ Age _____

Address _____

City, State/Province, ZIP/Postal Code_____

If licensed, how long? _____

Other hobbies_____

For ARRL use only	YMC
Edition 1 2 3 4 5 6 7 8 9 10 11 12	
Printing 1 2 3 4 5 6 7 8 9 10 11 12	

Occupation _____

From _____

_____

_____

EDITOR, YOUR MOBILE COMPANION
AMERICAN RADIO RELAY LEAGUE
225 MAIN ST
NEWINGTON CT 06111-1494

·· please fold and tape ··